50028

TD
884
.C48

The Chemistry of
air pollution

DATE DUE

DATE DUE

GAYLORD

PRINTED IN U.S.A.

The Chemistry of Air Pollution

Volume One in MSS' Series on Air Pollution

Papers by
Karl Westberg, Robert T. Cheng, Lyman
A. Ripperton et al.

MSS Information Corporation
655 Madison Avenue, New York, N.Y. 10021

Library of Congress Cataloging in Publication Data
Main entry under title:

The Chemistry of Air Pollution

(Air pollution, 1)
    1. Smog--Addresses, essays, lectures.
2. Automobile exhaust gas--Addresses, essays, lectures.
3. Photochemistry--Addresses, essays, lectures.
I. Westberg, Karl. II. Series: Air pollution
(New York), 1.
TD884.C48   628.5'3*2   73-11037
ISBN 0-8422-7152-X

TABLE OF CONTENTS

CREDITS AND ACKNOWLEDGEMENTS

Ajemian, R.S.; and N.E. Whitman, "Monitoring Carbon Monoxide in Ambient Air," *Journal of the Air Pollution Control Association*, 1970, 20:310-311.

Cheng, Robert T.; Morton Corn; and John O. Frohliger, "Contribution to the Reaction Kinetics of Water Soluble Aerosols and SO_2 in Air at PPM Concentrations," *Atmospheric Environment*, 1971, 5: 987-1008.

Cox, R.A.; and S.A. Penkett, "Oxidation of Atmospheric SO_2 by Products of the Ozone-Olefin Reaction," *Nature*, 1971, 230:321-322.

Del Vecchio, V.; P. Valori; C. Melchiorri; and A. Grella, "Polycyclic Aromatic Hydrocarbons from Gasoline-Engine and Liquefied Petroleum Gas Engine Exhausts," *Pure and Applied Chemistry*, 1970, 24: 739-748.

Dimitriades, Basil; B.H. Eccleston; and R.W. Hurn, "An Evaluation of the Fuel Factor through Direct Measurement of Photochemical Reactivity of Emissions," *Journal of the Air Pollution Control Association*, 1970, 20:150-160.

Dubois, L.; and J.L. Monkman, "Continuous Determination of Carbon Monoxide by Frontal Analysis," *Analytical Chemistry*, 1972, 44: 74-76.

Estefan, R.M.; E.M. Gause; and J.R. Rowlands, "Electron Spin Resonance and Optical Studies of the Interaction between NO_2 and Unsaturated Lipid Components," *Environmental Research*, 1970, 3:62-78.

Goldsmith, John R., "Carbon Monoxide Research — Recent and Remote," *Archives of Environmental Health*, 1970, 21:118-120.

——————————————— "Contribution of Motor Vehicle Exhaust, Industry, and Cigarette Smoking to Community Carbon Monoxide Exposures," *Annals of the New York Academy of Sciences*, 1970, 174: 122-134.

Judd, H.J., "Levels of Carbon Monoxide Recorded on Aircraft Flight Decks," *Aerospace Medicine*, 1971, 42:344-348.

Lee, Robert E., Jr.; Ronald K. Patterson; Walter L. Crider; and Jack Wagman, "Concentration and Particle Size Distribution of Particulate Emissions in Automobile Exhaust," *Atmospheric Environment*, 1971, 5:225-237.

McFarland, John H.; and C.S. Benton, "The Oxides of Nitrogen and Their Detection in Automotive Exhaust," *Journal of Chemical Education*, 1972, 49:21-24.

Reed, L.E.; and P.E. Trott, "Continuous Measurement of Carbon Monoxide in Streets, 1967-1969," *Atmospheric Environment*, 1971, 5:27-39.

Ripperton, Lyman A.; and Daniel Lillian, "The Effect of Water Vapor on Ozone Synthesis in the Photo-oxidation of Alpha-pinene," *Journal of the Air Pollution Control Association*, 1971, 21:629-635.

Schuchmann, H.P.; and K.J. Laidler, "Nitrogen Compounds other than NO_x in Automobile Exhaust Gas," *Journal of the Air Pollution Control Association*, 1972, 22:52-53.

Ter Haar, G.L.; D.L. Lenane; J.N. Hu; and M. Brandt, "Composition, Size and Control of Automotive Exhaust Particulates," *Journal of the Air Pollution Control Association*, 1972, 22:39-46.

Westberg, Karl; Norman Cohen; and K.W. Wilson, "Carbon Monoxide: Its Role in Photochemical Smog Formation," *Science*, 1971, 171:1013-1015.

PREFACE

In the United States air pollution reaches toxic thresholds in almost all urban areas. Its cost in terms of life and health is extreme. According to various governmental studies, the annual cost of damage resulting from polluted air is over $13 billion. If, for example, pollution could be cut by 50 percent, newborn babies would have an additional life expectancy of three to five years and deaths would be reduced 4.5 percent. Air pollution is also a factor in the development of diseases ranging from respiratory disorders to cirrhosis of the liver, and polluted air intensifies allergic reactions. Serious long-term research and conscientious political action are needed if these ominous trends are to be reversed.

Volume I in MSS' continuing series on air pollution present current research on the chemistry of smog. The composition of smog is discussed, and analytic techniques are described in detail. The role of carbon monoxide in smog formation is emphasized, and a special section of the book deals with the continuous measurement of carbon monoxide, both in ambient air and in areas where the concentration is particularly high. Motor vehicle emissions are discussed, and procedures for both analysis and control are presented.

The Chemistry of Smog

Carbon Monoxide: Its Role in Photochemical Smog Formation

KARL WESTBERG
NORMAN COHEN
K. W. WILSON

Until recently, the reactivity of carbon monoxide in polluted atmospheres has been all but ignored. Leighton (*1*) noted that the reaction of CO with O, H, or O_3 was demonstrably too slow to be significant. He did not discuss the reaction of CO with OH because, prior to 1966, the activation energy for that reaction was believed to be at least 7 kcal/mole (*2*), which would give a negligible reaction rate constant of approximately $10^5 M^{-1}$ sec^{-1} at atmospheric temperature. More recently, it has been demonstrated that the reaction of CO with OH (see reaction 1 below) has an activation energy of approximately 1 kcal/mole and that the rate constant at 25°C is $8 \times 10^7 M^{-1}$ sec^{-1} (*2*), thus making this reaction of possible significance in the atmosphere.

Recently Heicklen *et al.* proposed a mechanism to explain the previously unaccountably rapid rate of conversion of NO to NO_2 in polluted atmospheres (*3*). The mechanism involves three parallel chain reactions, each with OH as the chain carrier. Two chain reactions involve OH attack on hydrocarbons; the third involves the reaction of OH with CO and consists of the following steps (where M represents O_2 or N_2):

$$OH + CO \rightarrow CO_2 + H \qquad (1)$$
$$H + O_2 + M \rightarrow HO_2 + M \qquad (2)$$
$$HO_2 + NO \rightarrow OH + NO_2 \qquad (3)$$

The net reaction from these three steps is

$$CO + O_2 + NO \rightarrow CO_2 + NO_2 \qquad (4)$$

10

Experimental evidence is presented here for the effect of CO on the oxidation of NO in polluted atmospheres.

Experiments were performed in a 7.68-m^3, Teflon-lined, constant-temperature, stirred smog chamber (4). In each experiment approximately 3 parts per million (ppm) of isobutene, 1.5 ppm of NO or NO$_2$, and varying amounts of CO were mixed with an atmosphere of pure air and irradiated with light approximating the intensity and spectral characteristics of sunlight. Figures 1 and 2 show some of the results. A mixture consisting of NO, isobutene, carbon monoxide, and air was chosen for study because, in many ways, it simulates the chemical reactivity of a dilute mixture of automobile exhaust and air. The concentrations of hydrocarbon, NO, and CO used in these experiments are about an order of magnitude higher than those found in polluted atmospheres, but they are in roughly the correct proportions. High pollutant concentrations and the relatively high volume-to-surface ratio (V/S = 0.227 m) of the smog chamber helped to minimize the importance of wall reactions; identical results were obtained when the walls were newly cleaned or when they were well seasoned by numerous experiments.

As may be seen in Figs. 1 and 2, continuous measurements were made of the concentrations of O$_3$ [by means of an oxidant meter (Mast Development) with the use of Bufalini's method (5)], NO$_2$ (Saltzman reagent), NO + NO$_2$ (dichromate oxidation and Saltzman reagent), CO (infrared absorption), and isobutene (gas chomatography). Formaldehyde, peroxyacetyl nitrate (PAN), acetone, methyl nitrate, isobutene oxide, total hydrocarbon, and total oxidant were also monitored; for simplicity, these species have been omitted from the graphs since the effect of CO on their concentration was what would have been predicted from the altered O$_3$ concentration.

The air used in these experiments was purified by methods described elsewhere (4). Chemically pure grade isobutene, NO, and NO$_2$ were used without further purification. Carbon monoxide was purified by passage through a packed column cooled to liquid nitrogen temperatures to remove iron carbonyl, which is present in CO taken from conventional steel cylinders. In our CO samples the Fe(CO)$_5$ concentration was 0.15 percent. The purification procedure employed is effective only if carried out carefully, in which case it removes more than 99 percent of the iron carbonyl present. One of the experiments shown in Fig. 1 was performed with CO generated in a glass system by the reaction of H$_2$SO$_4$ with HCOOH. The results were identical to those obtained with CO taken from a steel cylinder and purified.

In Fig. 1 species concentrations are compared in the absence and in the presence of CO when the starting oxide of nitrogen is principally NO. In Fig. 2 the analogous curves for NO$_2$ are shown. These curves depict averages of several replications. Figure 1 demonstrates that CO markedly accelerates the disappearance of olefin, the conversion of NO to NO$_2$, and the appearance of ozone. However, the final ozone concentration is not noticeably altered. The effect of CO when NO$_2$ is the primary oxide of nitrogen is not so pronounced, as Fig. 2 indicates. The increase in the initial rate of ozone formation, although apparently slight as shown in Fig. 2, is not due to experimental uncertainty, but is a real effect. It actually represents an increase of about 50 percent in $d[O_3]/dt$ at the beginning of the experiment.

The detailed explanation of these

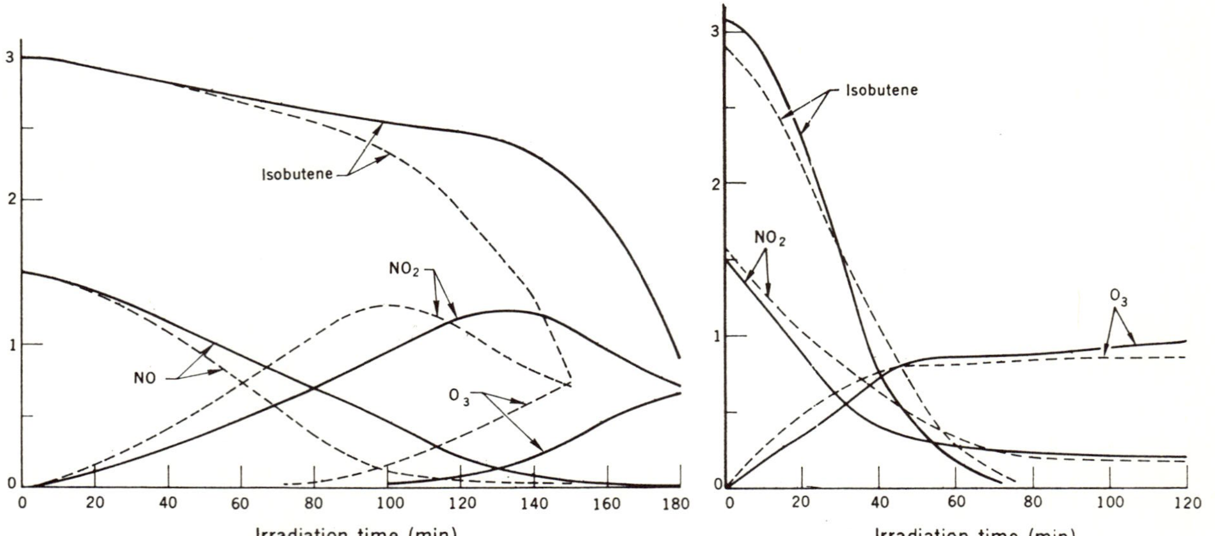

Fig. 1 (left). The effect of carbon monoxide on ozone formation. Curves shown are the results of photochemical experiments performed in an irradiated smog chamber starting with an atmosphere of air, 3 ppm of isobutene, 1.5 ppm of NO, less than 0.04 ppm of NO_2, and either 0 ppm of CO (solid lines) or 100 ppm of CO (dashed lines). The relative humidity was 70 percent; the temperature, 28°C. The NO, NO_2, and O_3 curves were corrected for the rapid reaction between NO and O_3 that occurred during the transfer of samples from the smog chamber to the measuring instruments. The results shown in this figure are averages of eight experiments. The experimental reproducibility is good: in matched pairs of experiments, the time required for the conversion of NO to NO_2 (defined as when $[NO] = [O_3]$) was always 31 ± 3 minutes shorter in the presence of added CO than in its absence. Nonreplicated measurements were made with different ratios of NO, CO, and isobutene, and carbon monoxide accelerated the conversion of NO to NO_2 in these experiments as well.   Fig. 2 (right). The effect of carbon monoxide on ozone formation. Curves shown are the results of experiments similar to those shown in Fig. 1, but with 1.5 ppm of NO_2 instead of NO. The difference in the ozone curves in the presence of either 0 ppm of CO (solid lines) or 70 ppm of CO (dashed lines) is explained in the text. We have no explanation for the corresponding differences in the isobutene and NO_2 curves; however, these curves were not nearly so reproducible as the O_3 curves.

experiments is subject to some speculation, but we believe that the following conclusions are established beyond reasonable doubt. First, it is apparent that CO is not inert in the production of photochemical smog, and smog chamber measurements made in the past in which CO was used as a tracer gas and its nonreactivity was assumed must be reevaluated. Second, and more important in the understanding of smog production, the experiments clearly implicate OH as a major chain carrier in the smog-producing system consisting of olefin, an oxide of nitrogen, CO, and air. There is no reaction other than reaction 1 by which CO plausibly can influence NO oxidation in this system. Finally, although we are not yet prepared to assess the magnitude of the role of CO in the real atmosphere, with its many different hydrocarbons, qualitatively it seems that the presence of CO will accelerate the early-morning conversion of NO to NO_2, thus hastening the appearance of oxidants.

The rate of ozone formation is closely controlled by the relative concentrations of NO and NO_2. Ozone is produced primarily by the photodissociation of NO_2 followed by the rapid combination of O atoms with O_2:

$$NO_2 + h\nu \rightarrow NO + O \qquad (5)$$
$$O + O_2 + M \rightarrow O_3 + M \qquad (6)$$

and is consumed principally by reaction with NO:

$$O_3 + NO \rightarrow NO_2 + O_2 \qquad (7)$$

Thus the ozone concentration increases in proportion to the amount of NO oxidized to NO_2. Quantitatively,

$$[O_3] \cong (k_5/k_7) [NO_2] / [NO] \qquad (8)$$

where k_5 is the product of the light absorbed (per unit time per unit volume) by NO_2 and its dissociation quantum efficiency and k_7 is the rate constant for reaction 7; in our system k_5/k_7 is $6.5 \times 10^{-10}M$. This relationship holds well at all times during the course of the smog experiments, even though, at times, either $[O_3]$ or $[NO]$ is too small to show on the graphs of Figs. 1 and 2.

The results included in Fig. 1 and in the first 40 minutes of the reaction in Fig. 2 show that CO accelerates the oxidation of NO to NO_2 and the rate of ozone formation. This result is what would be expected from reactions 1 through 3. However, as is shown in Fig. 2, when the ozone concentration becomes greater than about 0.6 ppm, and, hence, by reaction 8, $[NO_2]/[NO]$ is greater than 30, CO decelerates the further oxidation of NO to NO_2 and the rate of ozone formation. This behavior can be explained by a chain reaction consisting of reactions 1, 2, 9, and 10:

$$HO_2 + NO_2 \rightarrow HNO_2 + O_2 \qquad (9)$$
$$HNO_2 + h\nu \rightarrow OH + NO \qquad (10)$$

This chain reaction converts NO_2 to NO and will compete with reactions 1 through 3 under conditions of large $[NO_2]/[NO]$.

The carbon monoxide had 0.15 percent $Fe(CO)_5$ as its principal impurity (the hydrocarbon impurities were negligible, being less than 5 ppm). When iron carbonyl was not removed from the CO, the rate of oxidation of NO to NO_2 was accelerated even more than is shown in Fig. 1. In the presence of unpurified CO, only 60 minutes elapsed before all the NO was converted to NO_2 (defined as when $[NO] = [O_3]$). By comparison, with purified CO, 100 minutes elapsed (see Fig. 1); without CO, 130 minutes elapsed. Since the concentration of $Fe(CO)_5$ is an order of magnitude smaller than that of the NO, its pronounced effect can be explained only by a chain reaction. Because the carbonyl undergoes decomposition in sunlight (6) and is probably vulnerable to attack by oxygen atoms, a possible chain mechanism for its effect might include the following reactions:

$$Fe(CO)_5 + O \rightarrow Fe(CO)_4 + CO_2 \quad (11)$$

$$Fe(CO)_5 + \overset{h\nu}{\rightarrow} Fe(CO)_x + (5-x)CO, \\ x < 5 \quad (12)$$

$$Fe(CO)_x + O_2 \rightarrow Fe(CO)_xO_2 \quad (13)$$

$$Fe(CO)_xO_2 + (NO \text{ or } CO) \rightarrow \\ Fe(CO)_xO + (NO_2, CO_2) \quad (14)$$

$$Fe(CO)_xO + (CO \text{ or } NO) \rightarrow \\ Fe(CO)_x + (CO_2, NO_2) \quad (15)$$

Other reaction mechanisms may be equally possible, but reactions 11 through 15 do show that the observed effects of $Fe(CO)_5$ can be accounted for in a plausible manner. Iron carbonyl is highly reactive in other free-radical systems: trace amounts of $Fe(CO)_5$ either quench hydrocarbon-air flames or reduce their flame velocities substantially (7); this compound formerly was used as an antiknock agent in gasoline (6).

References and Notes

1. P. A. Leighton, *Photochemistry of Air Pollution* (Academic Press, New York, 1961), pp. 146–168, 183, 228.
2. D. L. Baulch, D. D. Drysdale, A. C. Lloyd, "High Temperature Reaction Rate Data No 1" (Department of Physical Chemistry, The University, Leeds, England, 1968).
3. J. Heicklen, K. Westberg, N. Cohen, in *Chemical Reactions in Urban Atmospheres*, C. S. Tuesday, Ed. (Elsevier, Amsterdam, in press).
4. G. J. Doyle, *Environ. Sci. Technol.* 4, 907 (1970).
5. J. J. Bufalini, *ibid.* 2, 703 (1968).
6. H. Remy, *Treatise on Inorganic Chemistry* (Elsevier, Amsterdam, 1956), vol. 2, p. 289; N. V. Sidgwick, *The Chemical Elements and Their Compounds* (Clarendon, Oxford, 1950), vol. 2, p. 1369; G. K. Rollefson and M. Burton, *Photochemistry and the Mechanisms of Chemical Reactions* (Prentice-Hall, New York, 1939), p. 362.
7. G. Lask and H. Gg. Wagner, *Symp. Combust.* 8, 433 (1962); W. J. Miller, *Combust. Flame* 13, 210 (1969).
8. We thank Prof. J. Heicklen and Drs. A. Bockian and S. W. Benson for valuable suggestions and advice, and I. Sauer and B. Kurtin for technical assistance in the experiments.

CONTRIBUTION TO THE REACTION KINETICS OF WATER SOLUBLE AEROSOLS AND SO₂ IN AIR AT PPM CONCENTRATIONS

ROBERT T. CHENG,* MORTON CORN and JOHN O. FROHLIGER

Abstract—A laboratory study of the heterogeneous catalysis of sulfur dioxide in the atmosphere involved the determination of the rate of atmospheric oxidation of sulfur dioxide to sulfuric acid in the presence of aerosols which were selected metal salts previously reported to act as catalysts. A new aerosol stabilizing technique was developed in which aerosol particles were deposited on inert supporting Teflon beads. The deposition was carried out in a fluidized-bed to ensure discrete aerosol deposition and to achieve a uniform distribution of aerosol concentration on the supporting beads. Because neither the physical shape of the aerosol nor their chemical properties were altered by the stabilizing process, kinetic data obtained in the laboratory by using stabilized aerosols can be extrapolated to describe reactions which may occur in the atmospheric environment.

Teflon beads with deposited aerosol particles were packed into a flow reactor. The catalytic oxidation of sulfur dioxide occurred within the reactor as sulfur dioxide gas in humid air mixture flowed through it. The reactor influent sulfur dioxide concentrations ranged from 3 to 18 ppm. Progress of the chemical reaction was continuously monitored by determining the sulfur dioxide concentrations in the reactor effluent with a microcoulometer. Reaction products were identified and quantified by analyzing the reactor contents after the completion of an experimental run.

It was observed that higher rates of sulfur dioxide oxidation were always associated with higher relative humidities of the air mixture. Because aerosol particles hydrated into solution drops at high relative humidities, the predominant mass transfer mechanism was absorption of sulfur dioxide by aqueous catalyst drops accompanied by chemical reaction in the liquid phase. Experimental results indicated that the over-all rate of reaction was controlled by the chemical reaction. Milligram for milligram, aerosols of $MnSO_4$, $MnCl_2$, and $CuSO_4$ were found to be 12.2, 3.5, and 2.4 times as effective as NaCl aerosol in promoting the oxidation of sulfur dioxide. $CuCl_2$ aerosol acted as a reactant in its reaction with sulfur dioxide; sulfur dioxide was oxidized to sulfuric acid while $CuCl_2$ was reduced CuCl. When $MnSO_4$ aerosol was used as catalyst, a reaction rate equation was derived which showed that the over-all reaction rate was first order with respect to the sulfur dioxide concentration in the gas phase. By extrapolating laboratory results to the industrial atmospheric environment, the rate of oxidation of sulfur dioxide in natural fog was estimated to be 2 per cent h^{-1}.

1. INTRODUCTION

SULFUR dioxide is a common atmospheric pollutant which arises mainly from combustion processes. Toxicologically, it is classified as a mild respiratory irritant, the main portion of which is absorbed in the upper respiratory tract. If the SO_2 present as an air pollutant remained unaltered until diluted, there is no evidence to suggest that it would cause adverse health effects in man at concentrations present in urban air pollution. However, studies of atmospheric chemistry have shown that SO_2 does not remain unaltered in the atmosphere but is converted to sulfuric acid, especially under conditions of high humidity and in the presence of particulate material. In terms of

* Supported by Training Grant No. AP 00041–03, National Air Pollution Control Administration, Environmental Protection Agency.

15

comparative toxicity, sulfuric acid in air is more irritating to the respiratory system than SO$_2$ in air, as demonstrated by studies using alterations in pulmonary function in guinea pigs as criteria (AMDUR, 1969). Therefore, the key to the role of SO$_2$ in air pollution toxicology lies not in the gas itself, but in its atmospheric chemistry.

Sulfur dioxide reacts very slowly with oxygen at 400°C to yield sulfur trioxide.

$$2 \, SO_2 + O_2 = 2SO_3 \quad \Delta H_{298} = -43{,}800 \text{ calorie} \tag{1}$$

Because the rate and subsequently the yield of reaction (1) is extremely slow at room temperature, the atmospheric oxidation of SO$_2$ must be due to other reaction processes. The most important ones are photochemical oxidation and catalytic oxidation. The photochemial oxidation of SO$_2$ occurs at a rate of 0.1 to 0.2 per cent h^{-1} (GERHARD *et al.*, 1955), while catalytic oxidation occurs at a faster rate and depends on the concentrations of the reactants and the type of catalyst used (JOHNSTONE and COUGHANOWR, 1958; JUNGE and RYAN, 1958). The atmospheric catalytic oxidation of SO$_2$ involves both water and dissolved oxygen, and requires the presence of a catalyst

$$2 \, SO_2 + 2 \, H_2O + O_2 \xrightarrow{\text{catalyst}} 2 \, H_2SO_4 \tag{2}$$

Known catalysts for the reaction include a number of metal salts, the more efficient of these being sulfates and chlorides of manganese and iron, which may exist in ambient air as suspended particulate matter. At high humidities, these particles can either act as condensation nuclei, or can undergo hydration to solution droplets. The oxidation process can properly be described as simultaneous absorption of two gases by liquid aerosols accompanied by chemical reactions in the liquid phase. Because this system of mass transfer with chemical reaction presents unique problems to experimental kineticists, this reaction system has not been extensively studied.

ANDERSON and JOHNSTONE (1955) absorbed SO$_2$ and oxygen in 0.01 molar manganese sulfate solution, using a Venturi atomizer for the absorber. The presence of the catalyst increased the rate of oxygen absorption several fold. However, the fluid mechanics of the system were too complex to estimate the kinetics of the reaction from absorption data.

JUNGE and RYAN (1958) bubbled SO$_2$ in air through dilute catalyst solutions. They found that, for the same catalyst, the maximum possible SO$_4^{2-}$ formation was a linear function of the SO$_2$ partial pressure in the air. Oxidation was rapid in the beginning of the experiment, but gradually reduced to zero rate. It was suggested that SO$_2$ oxidation in a catalyst solution is controlled by the pH of the solution. Sulfate formation decreases because the solubility of SO$_2$ decreases as the solution becomes more acid. Eventually, as a certain low pH is reached, SO$_4^{2-}$ formation ceases because SO$_2$ is exhausted from the liquid phase. Junge and Ryan's technique of using gas bubbles to simulate an aerosol–gas system was unrealistic because reaction kinetics could be drastically altered by mixing effects and by other mass transfer mechanisms.

JOHNSTONE and COUGHANOWR (1958) studied the simultaneous absorption and reaction of SO$_2$ and oxygen in a single drop of manganese sulfate solution. The surrounding gas was air with traces of SO$_2$ so that there was always excess oxygen at the interface. By employing the zero order rate constant obtained from previous experiments on homogeneous liquid phase oxidation of sulfurous acid, they were able to reduce the problem to one of gas diffusion into a liquid drop accompanied by zero order reaction in the liquid phase. A mathematical solution for the steady-state absorption of SO_2 by aqueous catalyst solution drops was developed. It was estimated that with one micron manganese sulfate crystals, the oxidation rate in fog droplets at 1 ppm SO_2 would be about 1 per cent min^{-1}.

JOHNSTONE and MOLL (1960) investigated the rates of catalytic oxidation of SO_2 to sulfuric acid in artificial fogs nucleated by small particles of manganese salt. Concentrations of acid up to 50 mg m^{-3} were formed in a few minutes at a relative humidity just below 100 per cent and initial SO_2 concentration of 250 ppm. These experimental results suffered from poor reproducibility because the maximum time for reaction was limited to a few minutes by the continuous evaporation, coagulation, and wall deposition of fog droplets.

MATTESON *et al.* (1969) investigated the oxidation of SO_2 in aqueous manganese sulfate aerosols. The reaction chamber was a plexiglas cylinder with a water jacket for temperature control. A mixture of SO_2, humid air, and aerosols of $MnSO_4$ continually flowed through the cylinder. Probe stations were installed at two points along the axis of the flow to facilitate sample withdrawal. Acid measured in samples collected on membrane filters was the indicator of oxidation. The reaction time was the time required for the gas to travel between the two probe stations. At the slowest gas flow rate (0.6 m^3 h^{-1}), the maximum reaction time was 15 min. Most of the experiments were characterized by reaction times between 1 and 5 min. Because hours of continuous sampling were required to obtain one aerosol sample with measurable acid formation, the sampling filters were replaced once every "few minutes" to avoid aerosol build up on the filter and to prevent reactions from occurring after the aerosols had deposited on the filter. If it is assumed that the "few minutes" reported by these authors means 5 min, then a liquid $MnSO_4$ aerosol collected early on the filter would still be in contact with the SO_2 atmosphere for 5 additional min beyond the recorded reaction time of 1–5 min. Expressed in another way, the reaction time for a portion of the collected aerosol was actually two to six times longer than the recorded reaction time. These investigators proposed a kinetic theory for the catalytic oxidation of SO_2 based on a four step chemical reaction involving the formation of intermediate complexes. However, it must be recognized that their theoretical treatment of kinetics is hypothetical because the experimental data obtained are basically incorrect.

In summary, reaction mechanisms and reaction kinetics relative to the atmospheric catalytic oxidation of SO_2 are in an early stage of development. Junge and Ryan's investigation of the oxidation of SO_2 in bulk catalyst solutions yielded valuable information on the effects of solution acidity on the rate of SO_2 oxidation. Johnstone and Coughanowr's method of exposing a single drop of catalyst solution to SO_2 in air was ingenious, even though they had to work with high SO_2 concentration (> 100 ppm) and large catalyst drops (> 700 μm). There were disagreements between Junge and Ryan's findings and Johnstone and Coughanowr's theory. For example, Junge and Ryan found that the rate of SO_2 oxidation was of first order with respect to SO_2 concentration in the gas phase, but Johnstone and Coughanowr used a zero order reaction mechanism to derive their theoretical equations. Attempts by Johnstone and Moll and by Matteson *et al.* to measure the reaction between SO_2 and catalyst aerosol in an air-dispersed environment have failed largely due to the unstable nature of the aerosol phase. The lack of verifying experimental kinetic data also hampered the development of kinetic theory concerning this catalytic reaction.

17

In the study reported here, a new experimental method was developed to provide mechanical and thermodynamic equilibrium to the aerosol–gas system without altering the physical shape and chemical properties of the aerosol (CHENG, FROHLIGER and CORN, 1971). By using this method, kinetic data obtained in the laboratory can be extrapolated to describe reactions which may occur in the atmosphere.

2. EXPERIMENTAL METHODS

The heart of the experimental apparatus consisted of a packed bed of beads coated with aerosol particles of each of the potential catalysts investigated. Details relative to the preparation of the coated beads and a discussion of their use to simulate atmospheric gas–aerosol reactions have been presented elsewhere (CHENG, FROHLIGER and CORN, 1971).

Basically, this new technique comprised the deposition of catalyst aerosol particles on an inert substrate such as Teflon beads. The deposition process was carried out in a fluidized-bed system so that both the uniform distribution of the individual aerosol particles on the substrate, and the uniform distribution of the aerosol concentration throughout the substrate was simultaneously achieved. Teflon beads were chosen as the aerosol supporting substrate because they are chemically inert to SO_2. An additional advantage of Teflon is its low specific surface free energy. At high relative humidity, the hydrated catalyst aerosol particles did not spread on the Teflon surface, but maintained themselves as individual spherical drops, thus simulating the behavior of hydrated particles in the atmosphere.

Micron and submicron size aerosol particles of $MnSO_4$, $MnCl_2$, $NaCl$, $CuSO_4$ and $CuCl_2$ were deposited on Teflon beads by the aerosol stabilizing technique. The beads with deposited aerosol particles were packed into a flow reactor which was used to obtain kinetic data (FIG. 1).

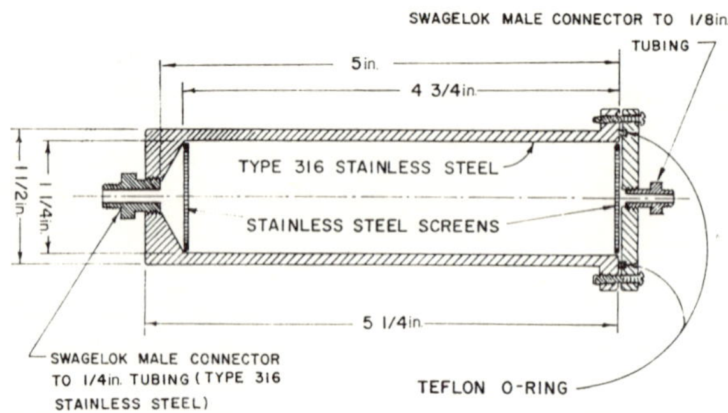

FIG. 1. Details of the experimental flow reactor.

Catalytic oxidation of SO_2 occurred within the reactor as a mixture of SO_2 and humid air flowed through it. Progress of the chemical reaction was continuously monitored by determining the reactor effluent SO_2 concentration with a microcoulometric titrating system. Some reaction products were identified and quantified by analyzing the reactor contents after the completion of an experimental run. The experimental system is shown schematically in Fig. 2. Components of the system and procedures used will be described in detail.

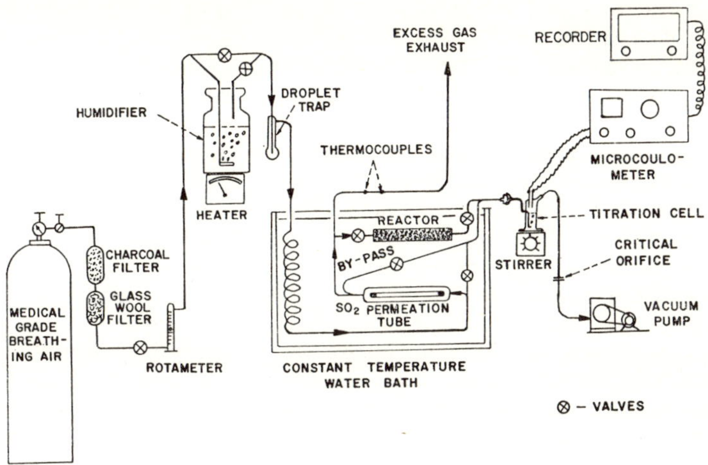

Fig. 2. Schematic diagram of the experimental system.

(a) *Air supply apparatus*

Pressurized medical grade breathing air was used for dilution air in all experiments. The air was cleaned of contaminants by passing it through an activated charcoal filter and a glass wool filter. Flow rate was adjusted by regulating and metering devices.

(b) *Humidity adjustment and measurement*

Moisture was added to the air stream as it flowed through a bubbler type humidifier. Moisture content of the air stream was adjusted by regulating the water temperature of the humidifier, or by controlling the amount of air flowing to the bypass device. Relative humidity was measured with a combination of dry bulb and wet bulb thermo-couples. They were made from fine copper and constantan wires and were housed in a length of 1/8 in. i.d. lucite tube which was inserted in the excess gas exhausting line. A Leeds & Northrup Model 8662 Precision Potentiometer was used to measure the thermocouple e.m.f.

(c) *Sulfur dioxide source and temperature control*

Teflon permeation tubes were used for preparing the low concentrations of SO_2 in air required in these studies.* O'Keeffe and Ortman proposed the use of Teflon permeation tubes for the preparation of primary and working standards of trace quantities of volatile atmospheric pollutants (O'Keeffe and Ortman, 1966). Scaringelli later described calibration procedures for these devices (Scaringelli *et al.*, 1967). We used a constant temperature water bath to maintain the permeation tubes at the desired temperature $\pm$ 0.1°C. Both the SO_2 permeation tube holder and the reactor were immersed inside the bath.

* Permeation tubes were purchased from Analytical Instrument Development, Inc., West Chester, Pennsylvania.

(d) *Reactor and gas feed rate control*

The reactor was constructed from a $1\frac{1}{4}$ in. i.d. type 316 stainless steel tube, was 5 in. long and had a gross volume of about 100 cm³ (FIG. 1). A total of 64 g of Teflon beads could be packed into this reactor. A section of the reactor close to the gas entrance was packed with pure Teflon beads (without carrying catalyst aerosols) to establish a more uniform gas flow pattern within the reactor; the remaining reactor volume was packed with catalyst bearing beads. By varying the length of the section packed with the pure Teflon beads, the remaining reactor volume was also varied. Consequently, we obtained various effective reactor volumes by working with the same reactor. All tubings, fittings, and valves in the reactor system were made from type 316 stainless steel. Soldering joints were not exposed to the gas flow because silver solder can react with SO_2 at high relative humidity. Gas flow rate through the reactor was controlled by a critical orifice downstream from the SO_2 detector. The reactor influent SO_2 concentration could be regulated to a predetermined level either by using the same critical orifice but varying the air flow rate through the SO_2 permeation tube, or by employing different critical orifices to establish various rates of air flow through the permeation tube. Gas flow inside the reactor was laminar with particle Reynolds number around 1×10^{-4}. A uniform flow distribution pattern across the reactor axis was assumed because the ratio of reactor diameter to the diameter of the Teflon beads was over 40 (SCHWARTZ and SMITH, 1953). There were four flow directing valves around the reactor. By using the right combination of these valves one could bypass the reactor to permit either pure air or SO_2 and air mixture to enter the coulometric titration cell.

(e) *Sulfur dioxide measurement*

The reactor influent and effluent SO_2 concentrations were measured by a coulometric titration system (Dohrmann Model C-200 Microcoulometer with T-300-P titration cell). The microcoulometer continuously titrated to a potentiometric endpoint the SO_2 collected from the gas stream. The electrolyte in the titration cell consisted of a solution containing 0.04 per cent acetic acid and 0.05 per cent potassium iodide in water. The amount of current required to generate enough iodine for the reaction

$$I_2 + SO_2 + 2H_2O \longrightarrow H_2SO_4 + 2HI \qquad (3)$$

was measured by the voltage drop across precision resistors and was recorded continuously by means of a 1-mV recorder. The instrument sensitivity rate was 30 ng s^{-1}

(f) *Procedures for a standard run*

After the reactor was properly cleaned with concentrated nitric acid, dried and packed, an air stream with the desired moisture content but without SO_2 was delivered to the reactor for 10 min. This procedure allowed the catalyst aerosols to be preconditioned at the relative humidity to be used in the experiment. At zero time, by means of bypass valves, air containing known concentrations of SO_2 and water was permitted to enter the reactor at a constant flow rate.

(g) *Calibration of the experimental reactor*

A reactor packed fully with bare Teflon beads (no catalyst), was designed as a dummy reactor. In a dummy reactor, the effluent SO_2 concentration rose to approach the level of influent SO_2 concentration a short time after the gas mixture entered the reactor. Most of this time lag was due to replacement of the reactor volume by the influent gas, while a minor part might have resulted from reaction or adsorption of SO_2 on reactor walls and surfaces of beads. When the experimental relative humidity was below about 40 per cent, the effluent SO_2 concentration equaled the influent SO_2 concentration within 4 min after the start of the experiment. This rapid equilibration occurred for all influent SO_2 concentrations from 3 to 18 ppm, indicating SO_2 did not interact (adsorption or reaction) with the dummy reactor. At higher relative humidities a small portion of the influent SO_2 was adsorbed or absorbed by the moisture films covering the bead and reactor surfaces. FIGURE 3 shows the time dependence of reactor effluent SO_2 concentration for four dummy reactors at 0, 36, 80, and 95 per cent relative humidities. At 0 and 36 per cent relative humidities, it took about 4 min for the effluent SO_2 concentration to reach the influent SO_2 concentration, while the same equilibration required 35 min at 95 per cent relative humidity. By comparing the SO_2 break-through curve of a reactor with catalyst beads to that of a dummy reactor under identical experimental conditions (temperature, relative humidity, influent SO_2 concentration, gas flow rate, etc.), we assessed the reduction in effluent SO_2 concentration attributable to catalyst aerosols. The curves of SO_2 concentration vs. time for repeat runs with a dummy reactor at specified operating conditions were essentially duplicates of each other and were virtually indistinguishable one from the other.

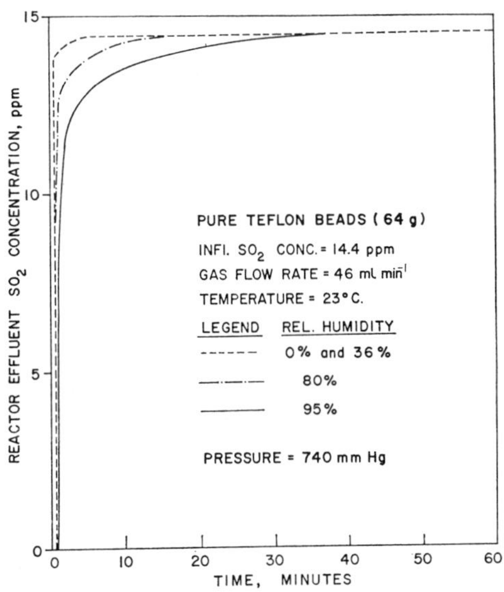

FIG. 3. Reactor effluent SO_2 concentration vs. time for dummy reactors at 14.4 ppm influent SO_2 concentration.

(h) *Determination of sulfate concentration*

Because SO_2 was converted predominantly into sulfuric acid or sulfates, an analysis of the reactor contents for sulfate served to check the sulfur balance. For selected experimental runs, Teflon beads were removed from the reactor after completion of the run and sulfate samples were prepared by washing and redissolving the newly formed sulfates on the Teflon beads. The same washing and redissolving technique was also incorporated into procedures for preparing a sulfate calibration curve. The Turbidimetric Barium Sulfate Method (ENVIRONMENTAL HEALTH SERIES AIR POLLUTION 999-AP-11, 1965) was used to determine sulfate concentrations.

(i) *Selection of catalysts*

Five kinds of metal salts were used for catalysts: sodium chloride (NaCl), manganese chloride ($MnCl_2$), manganese sulfate ($MnSO_4$), copper chloride ($CuCl_2$), and copper sulfate ($CuSO_4$). Sodium chloride in fine particle form is produced in nature by evaporation of sea spray. The geometric mean concentrations for both copper and manganese in suspended particulate matter at urban locations in the U.S. are about 0.1 μg m^{-3}; maximum concentrations for each are 10 μg m^{-3} (DEPT. OF HEALTH, EDUCATION, AND WELFARE, DIV. OF AIR POLLUTION, 1962, 1966).

The use of copper salts as catalyst for the oxidation of sulfurous acid has been noted (JUNGE and RYAN, 1958). In addition, the effect of cupric ions on the oxidation of sodium sulfite has been widely investigated and its reaction mechanism partially clarified. Therefore, two copper salts were selected for study in the hope that information on sodium sulfite oxidation could be applied to SO_2 oxidation.

3. RESULTS AND DISCUSSION

(a) *Fluid flow characteristics in the reactor*

At 46 ml min^{-1} gas flow rate a pulse injection of SO_2 was given to the reactor. FIGURE 4 presents reactor response to this pulse by showing concentration in the exit stream of the reactor as a function of time. The mean residence time of a fluid element in the reactor was 1.7 min; the variance of residence time was 0.17.

Consider the plug flow of a fluid, on top of which is superimposed some degree of backmixing or intermixing, the magnitude of which is independent of position within the reactor. This condition of dispersed plug flow implies that stagnant pockets and gross bypassing or shortcircuiting of fluid did not exist in the reactor. For a pulse tracer input into fluid in dispersed plug flow, the parameter which correctly characterizes the role played by dispersion is the reactor dispersion number (D/uL). It varies from zero for plug flow to infinity for backmix flow. Individually, D is called the axial dispersion coefficient $(L^2 T^{-1})$, u the fluid velocity (LT^{-1}), and L the length of the reactor (L). The reactor dispersion number (D/uL) is closely related to the mean and variance of the residence time distribution (VAN DER LAAN, 1958). The reactor dispersion number for our experimental reactor was 0.03, indicating a small but significant deviation from plug flow conditions.

For multiple reactions of the consecutive type TICHACEK (1963) presented equations accounting for the effect of axial dispersion and showed that for systems with small but significant deviations from plug flow, say for $(D/uL) = 0.05$, the fractional decrease in the maximum amount of reaction products formed is closely approximated by the value of (D/uL) itself. Therefore, for $(D/uL) = 0.03$, one can expect the conversion in an ideal plug flow reactor to be approximately 3 per cent larger than the

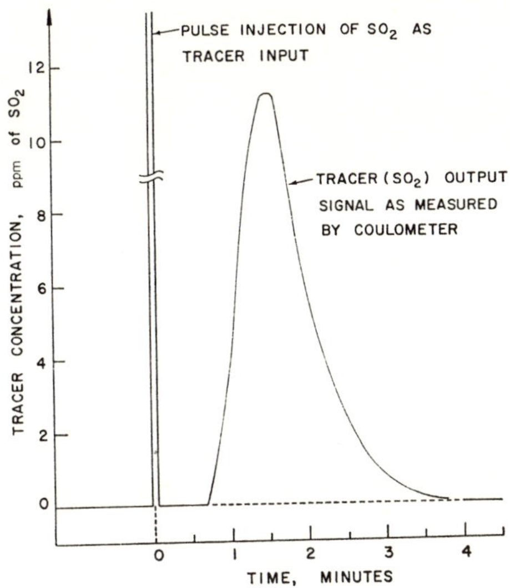

FIG. 4. Distribution of residence time of fluid in the experimental reactor, measured following an upstream pulse injection of sulfur dioxide.

conversion in the experimental reactor. To transform chemical conversion obtained with this new experimental reactor to that which would occur in an ideal plug flow reactor, one simply multiplies the former value by a factor of 1.03.

(b) *Reaction with manganese sulfate as catalyst*

Duplicate runs using the reactor with catalyst coated Teflon beads agreed to within 3 per cent, using reactor effluent SO_2 concentration at the same reaction time as the criterion for duplication. FIGURE 5 shows the SO_2 break-through curve of an experimental run using 15.5 g of $MnSO_4$ aerosol coated beads as the catalyst. The aerosol concentration on these beads was 0.033 mg of $MnSO_4$ per gram of beads; thus, the reactor contained a total of 0.51 mg of $MnSO_4$ as catalyst. The dotted curve represents the SO_2 break-through curve of the dummy reactor for the experimental conditions indicated. By subtracting the SO_2 concentration of the catalyst reactor at any time from the SO_2 concentration of the dummy reactor at the same time, we obtained the amount of SO_2 removed from the influent stream due specifically to the presence of catalyst in the reactor. In FIG. 6 the fraction conversion X is the ratio of moles of SO_2 removed in the reactor divided by the moles of SO_2 delivered to the reactor. Under unsteady state conditions the SO_2 removed may not be entirely converted to sulfur trioxide or sulfuric acid at any specified reaction time; it is converted only in the sense that all SO_2 removed is either adsorbed or absorbed or reacted by the catalyst. Under steady-state conditions, however, the rate of adsorption or absorption is in equilibrium with the rate of reaction, and the SO_2 removed is truly converted by chemical reaction.

FIGURE 6 demonstrates that the progress of $MnSO_4$ catalyzed oxidation of SO_2 can be divided into three stages. The conversion was proceeding at maximum rate during

23

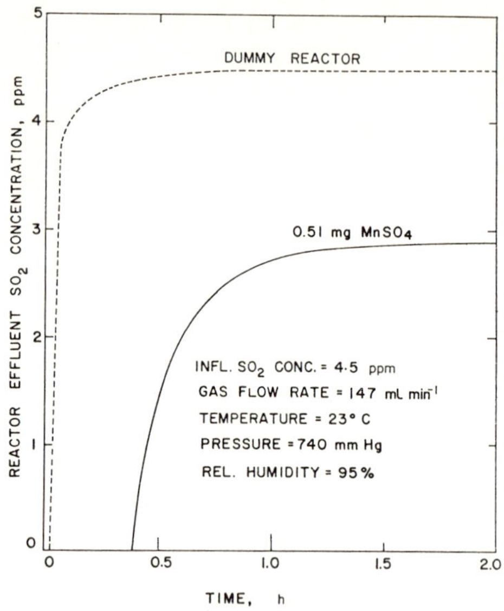

FIG. 5. Sulfur dioxide break-through curve of an experimental run using 0.51 mg of $MnSO_4$ aerosol as catalyst (influent SO_2 concentration 4.5 ppm).

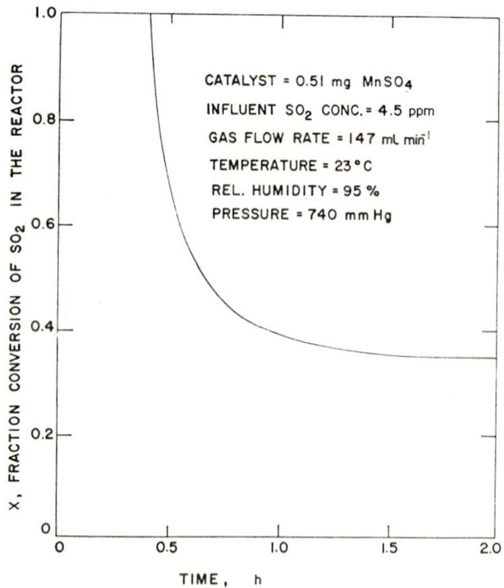

FIG. 6. Fraction conversion of SO_2 in the reactor vs. time for 0.51 mg $MnSO_4$ aerosol at 4.5 ppm influent SO_2 concentration.

the initial 23 min of the reaction. All of the influent SO_2 was converted and SO_2 did not appear at the exit end of the reactor. A transitional period (from 23 min to about one and one-half hours) followed which was characterized by decrease of the SO_2 conversion rate from the initial maximum value to a certain steady value. From then on, reaction occurred in the third stage, essentially steady-state conversion of SO_2 in the reactor. This three stage rate process can be related to the change of solubility of SO_2 in water solution as the solution becomes more acidic.

(i) *Solubility of sulfur dioxide in water and in sulfuric acid.* The absorption of atmospheric SO_2 by water results in the formation of sulfurous acid which in turn is ionized in the solution.

The observed initial rapid conversion of SO_2 in the reactor would appear to be due to rapid solution of gas phase SO_2 into the catalyst solution drops. The diffusion of the ionized species (HSO_3^-, SO_3^{2-}) contributed to the total diffusion of SO_2. Thus, if the rate of reaction was considerably smaller than the rate of absorption, a large amount of SO_2 would be removed from the reactor during the initial stage of unsteady state conversion.

But this initial stage of rapid conversion would soon be affected by the increase of sulfuric acid in the solution drops. Because sulfuric acid in dilute concentration undergoes complete dissociation

$$H_2SO_4 \underset{\longleftarrow}{\overset{K}{\longrightarrow}} HSO_4^- + H^+ \qquad K > 10 \qquad (4)$$

the added hydrogen ion concentration would diminish the solubility of SO_2. This period of rapid decrease in solubility with increasing concentration of H_2SO_4 was manifested as the transitional state in reactor conversion. Finally, as the solution acid concentration exceeded a certain level, the high hydrogen ion concentration prevented further dissociation of H_2SO_4 and all the dissolved SO_2 existed as unionized sulfurous acid.

The solubility of SO_2 in sulfuric acid solutions was calculated by COUGHANOWR (1956) at 25°C for a partial pressure of SO_2 of 0.0013 atm. (FIG. 7) and is in excellent agreement with the experimental points reported by JOHNSTONE and LEPPLA (1934) for the same conditions. The solubility of SO_2 first decreases rapidly with increasing concentrations of H_2SO_4, but when the concentration of sulfuric acid reaches about 0.1 N, the solubility becomes nearly constant and equal to the calculated value, which is based on the assumption that all the dissolved SO_2 is in the unionized form. Under this condition, one molecule of SO_2 oxidized in the catalyst solution drops will soon be replaced by the dissolving of another molecule of SO_2 from the gas phase so that the saturation solubility value can be maintained. In other words, the rate of chemical conversion becomes equal to the rate of gas absorption. If the chemical reaction itself approaches steady state (i.e., the rate is not affected by the accumulation of reaction products and the catalyst is not poisoned), then the over-all conversion should also appear as a steady state process. The apparently constant rate of conversion during the third stage of the fraction conversion vs. time curve (FIG. 6) confirms this theory.

(ii) *Mechanism of the reaction.* BASSETT and PARKER (1951) proposed that in the oxidation of sulfurous acid in catalyst solutions the active catalyst is a transitional metal-sulfite complex. The complex forms, picks up a molecule of oxygen, and rearranges to form either a sulfite or a dithionate group. The resulting rearranged

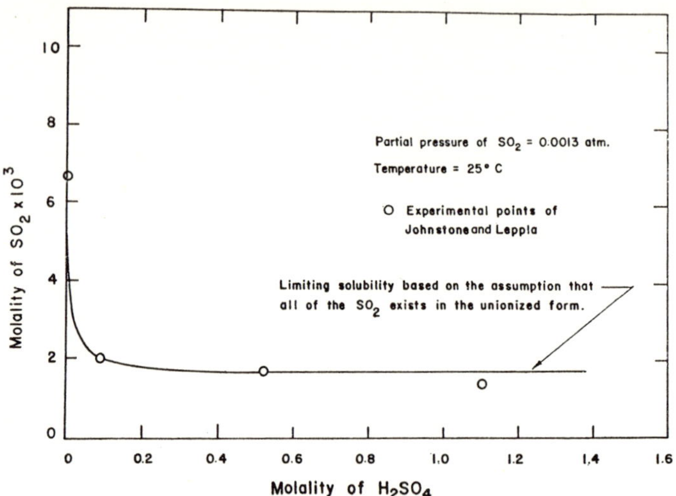

Fig. 7. Solubility of SO_2 in H_2SO_4 solution at a constant pressure of SO_2 (Coughanowr).

complexes are unstable and form new sulfite complexes and either sulfuric acid or dithionic acid. Pritchett (1961) observed an accelerating rate of oxidation of sulfurous acid catalyzed by manganese sulfate. He proposed a mechanism which involved the formation of several intermediate compounds and suggested that the reaction chain started with the combination of sulfite ion and metal salt in solution. The first step of the proposed mechanism is

$$MnSO_4 + SO_3^{2-} \rightleftharpoons P \tag{5}$$

where P is an unidentified intermediate compound. However, results from this investigation and the above discussion on the solubility of SO_2 indicate that during the region of steady-state reaction, the dissolved SO_2 exists entirely as unionized sulfurous acid. Essentially, bisulfite ion (HSO_3^-) does not exist in the solution, let alone sulfite ions (SO_3^{2-}) whose existence in solution depends on the rather weak dissociation of bisulfite ion ($K_2 = 1 \times 10^{-7}$). Therefore, the logical initial step of the reaction should involve the combination of $[SO_2]_{aq}$ with metal ions to form an intermediate complex. Results from work reported here suggests that the initial step might be

$$x\, SO_2 + Mn^{2+} \rightleftharpoons Mn \cdot x\, SO_2^{2+} \tag{6}$$

(iii) *Kinetics of the reaction.* When the reactor conversion is high, the reaction rate changes from one end of the reactor to the other. The system can be treated as an integral steady-state reactor (Levenspiel, 1962), as described by equation (7).

$$-R = \frac{dX}{d\,(W/F_0)} \tag{7}$$

where X = fraction of reactant converted

R = rate of reaction [moles reacted (time)$^{-1}$ (mass of catalyst)$^{-1}$]

W = mass of catalyst

F_0 = reactor input (moles time^{-1})

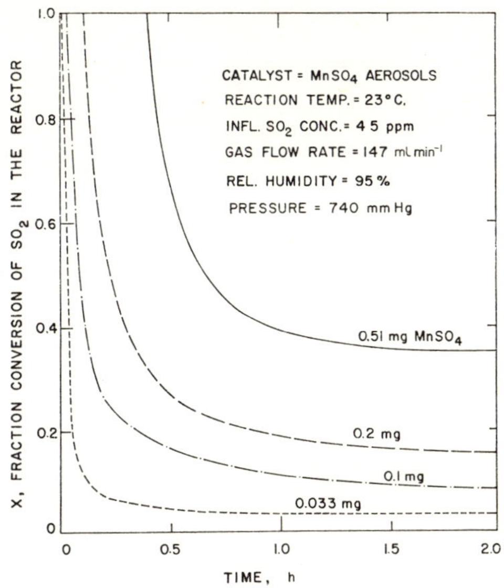

FIG. 8. Fraction conversion of SO_2 in the reactor vs. time for various weights of $MnSO_4$ aerosol.

In the study reported here four experiments were performed with SO_2 in air at 23°C temperature and 95 per cent relative humidity; the gas feed rate was 1.74 μg of SO_2 min⁻¹. The concentration of catalyst aerosol in the Teflon beads was 0.033 mg $MnSO_4$ per gram of beads. The total weight of manganese sulfate aerosol in the reactor ranged from 0.033 to 0.51 mg. After about two hours of reaction time each run reached steady-state conversion (Figure 8). The reaction rate equation which describes these results is

$$-R \left(\frac{\mu\text{g of SO}_2}{(\text{min}) \, (\text{mg of MnSO}_4)} \right)$$

$$= 1.33 \times 10^{-4} \left(\frac{\text{m}^3 \text{ of air}}{(\text{min}) \, (\text{mg of MnSO}_4)} \right) C \left(\frac{\mu\text{g of SO}_2}{\text{m}^3 \text{ of air}} \right) \tag{8}$$

(c) *Effects of gas phase diffusional resistance*

The effects of gas phase diffusional resistance was evaluated. The amount of chemical conversion was measured at a value of W/F (weight of catalyst divided by gas feed rate) at which the gas flow rate was low and again at a higher gas flow rate, but at the same W/F ratio. If the two values of conversion coincide, the effect of gas phase diffusion is negligible. If the conversions are different, the resistance offered by gas phase diffusion could play an important role in determining the over-all rate of reaction. In this study, three experiments were performed with $MnCl_2$ aerosols at the same influent SO_2 concentration of 3.3 ppm; air temperature was 23°C and air relative humidity was 98 per cent. Different critical orifices were used downstream from the reactor to regulate the flow rate of the gas mixture through the reactor to 147 ml min⁻¹, 83 ml min⁻¹, and 46 ml min⁻¹. The corresponding gas feed rates were 1.28, 0.725, and

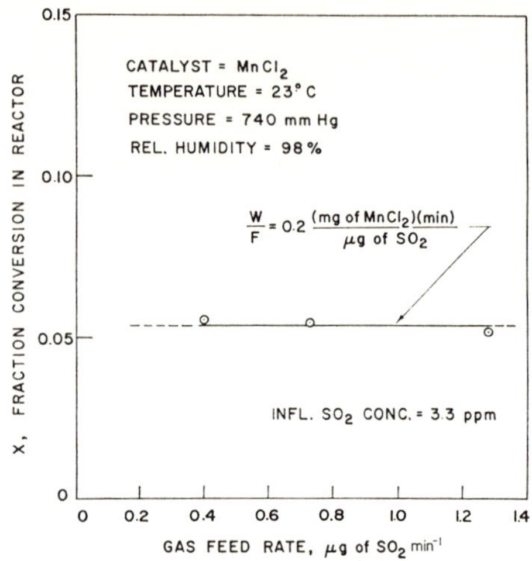

FIG. 9. Conversion vs. gas feed rate; test for influence of gas phase resistance.

$0.40 \ \mu g$ of SO_2 min^{-1}; the W/F ratio was maintained at a constant value of 0.2 (mg of $MnCl_2$) (min) (μg of SO_2)$^{-1}$ by varying the total weight of $MnCl_2$ in the reactor. All three runs appeared to have reached steady-state conversion after 60 min. FIGURE 9 indicates the steady-state fraction conversions for these runs. It was concluded that gas phase diffusional resistance did not influence the over-all rate of conversion.

(d) *Reaction with other metal salts as Catalysts*

When the reactor was packed with stabilized aerosols of sodium chloride, manganese chloride, or copper sulfate, the reactor SO_2 break-through curves exhibited characteristics similar to those obtained when the reactor contained manganese sulfate aerosol on Teflon beads. The progress of SO_2 conversion can again be divided into an initial stage with very rapid rate of conversion, followed by a transitional stage with decreasing rate of conversion, followed by a final stage of steady state conversion. Each salt differed in effectiveness as a catalyst for the oxidation of sulfur dioxide. Table 1 indicates the steady-state conversions achieved with these catalysts under non-uniform experimental conditions. Because fraction conversion is proportional to residence time, but is essentially independent of reactor influent SO_2 concentration, $MnSO_4$ was the most effective, and NaCl the least effective, of the four catalysts tested. If the catalytic strength of 0.36 mg NaCl aerosol at 1.7 min reactor mean residence time is assigned a factor of effectiveness of 1.0, then the effectiveness factors for the same weights of $CuSO_4$, $MnCl_2$ and $MnSO_4$ are 2.4, 3.5 and 12.2, respectively. These results confirm findings by ADMUR and UNDERHILL (1968), who used a biological parameter of response and concluded that sodium chloride is a poor catalyst for the SO_2 to SO_4^{2-} conversion.

28

(e) *Effects of liquid phase diffusional resistance*

The individual steps in the catalytic oxidation of sulfur dioxide in aqueous catalyst are:
(1) Diffusion of SO_2 to the outside of the solution drops.
(2) Diffusion of SO_2 from the surface of the drops to the interior.
(3) The catalytic reaction itself.

TABLE 1. EFFECT OF CATALYST ON SO_2 OXIDATION

Catalyst	Weight (mg)	Mean residence time (min)	Influent SO_2 conc. (ppm)	Fraction conversion	Effectiveness factor*
NaCl	0.36	1.7	14.4	0.069	1.0
$CuSO_4$	0.15	1.7	14.4	0.068	2.4
$MnCl_2$	0.255	0.52	3.3	0.052	3.5
$MnSO_4$	0.51	0.52	3.3	0.365	12.2

* The catalytic effectiveness of the various materials was compared to that of NaCl. Thus, Effectiveness factor = 1.0 × (ratio of weight of catalyst in the reactor) × (ratio of reactor mean residence time) × (ratio of reaction conversion of SO_2 in the reactor).
The effectiveness factor for $MnSO_4$, for example, is:

$$1.0 \times \frac{0.36}{0.51} \times \frac{1.7}{0.52} \times \frac{0.365}{0.069} = 12.2.$$

We have shown that gas phase diffusion of SO_2 did not influence the over-all rate of conversion. If the liquid phase diffusion was faster than the rate of chemical reaction, the reactants would diffuse into the innermost parts of the catalyst drop and the entire drop would have a concentration of SO_2 approximately equal to the SO_2 concentration at the drop surface. However, if the chemical reaction proceeded more rapidly than diffusion, SO_2 concentration would be altered before diffusing very far into the drop. The reaction would occur mainly in the outer periphery of the drop and part of the interior would not be utilized.

JOHNSTONE and COUGHANOWR (1958) suggested that the over-all rate of absorption of SO_2 was controlled by liquid phase diffusion. On the other hand, MATTESON *et al.* (1969) suggested that the over-all rate of the catalytic oxidation of SO_2 was controlled by a four-step chemical reaction involving the formation of intermediate complexes.

In this study the concentrations of catalyst salts in the solution drops were approximately the same for all salts studied. Therefore, the liquid phase diffusivity of SO_2 remained essentially constant regardless of the salt utilized as catalyst. If the over-all SO_2 conversion rate was controlled by liquid phase diffusion, then the conversion would have been independent of the catalyst salt used. However, a twelve-fold increase in conversion was achieved by changing from NaCl to $MnSO_4$ catalyst. These results suggest that the rate controlling mechanism is within the chemical reaction itself.

(f) *Reaction with copper chloride aerosol*

When the reactor was packed with stabilized cupric chloride ($CuCl_2$) aerosols, the SO_2 break-through curve appeared entirely different (FIG. 10). A general trend of

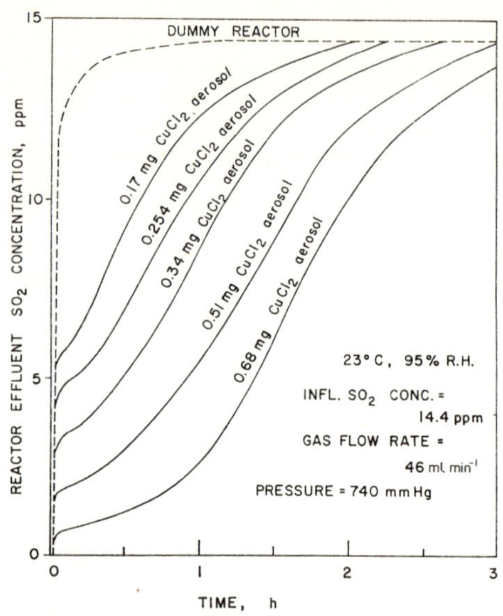

FIG. 10. Sulfur dioxide break-through curves for runs with various amounts of $CuCl_2$ aerosols at 14.4 ppm influent SO_2 concentrations.

decreasing conversion with increasing reaction time is indicated. Steady-state conversion was never reached; the reactor effluent SO_2 concentrations equalled the influent SO_2 concentrations a few hours after the start of an experiment. An SO_2 break-through curve joins the dummy reactor curve when zero conversion is reached. The area bounded by the dummy reactor curve and an SO_2 break-through curve is interpreted as representing the amount of SO_2 removed in the reactor because of the presence of cupric chloride aerosol. A plot of SO_2 removed in each run vs. the weight of $CuCl_2$ aerosol used, yielded a straight line (FIG. 11). Approximately one micromole of SO_2 was removed for every two micromoles of $CuCl_2$ in the reactor.

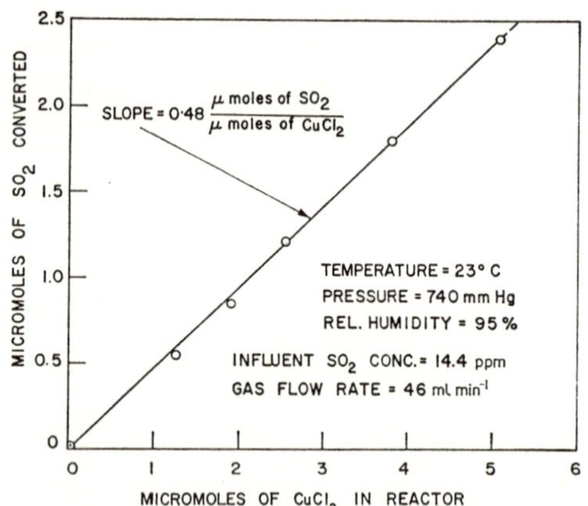

FIG. 11. Micromoles of SO_2 conversion vs. micromoles of $CuCl_2$ in reactor.

When we consider the shape of the SO_2 break-through curves, the linear proportionality between weight of $CuCl_2$ and quantity of SO_2 removed, and the stoichiometric relationship between SO_2 and $CuCl_2$, it seems unlikely that $CuCl_2$ aerosol acted as catalyst in its reaction with SO_2. We conclude that $CuCl_2$ was probably reacting with SO_2. We know that $CuCl_2$ was not transformed into $CuSO_4$ in this reaction because $CuSO_4$ has been shown to be an effective catalyst and would further promote the oxidation of SO_2 if present in the drops. Analyses for sulfate in the reactor bed after the completion of experimental runs showed that SO_2 was predominantly oxidized to sulfates. If SO_4^{2-} was not combined with copper in the form of $CuSO_4$, then it must have existed in the solution drops as H_2SO_4 or as some stable complex involving H_2SO_4 and $CuCl$. An oxidation–reduction reaction that produces sulfuric acid and also fits the stoichiometric relationship of these results is:

$$SO_2 + 2\,CuCl_2 + 2\,H_2O \rightleftharpoons 2\,CuCl + H_2SO_4 + 2\,HCl \qquad (9)$$

SO_2 is oxidized into sulfuric acid while copper is reduced from the cupric form to the cuprous form.

(g) *Sulfate analysis*

TABLE 2 shows the results of sulfate analysis for experiments with $CuCl_2$ and $MnCl_2$ aerosols. The conversion from SO_2 to SO_4^{2-} was consistently less than 100 per cent. Incomplete recovery of all SO_4^{2-} from Teflon bead samples during the washing process could explain this result. An alternate possibility is that all of the removed SO_2 was not converted into sulfates or sulfuric acid. The principle product of the air oxidation of sulfurous acid is sulfuric acid, but some dithionic acid is also formed (BASSETT and HENRY, 1935).

$$2\,H_2SO_3 + \tfrac{1}{2}\,O_2 \xrightarrow{\text{catalyst}} H_2S_2O_6 + H_2O \qquad (10)$$

This reaction may have accounted for a portion of the removed SO_2 not being converted to SO_4^{2-}.

(h) *Effects of relative humidity*

Even at low relative humidities, an aqueous film is found on the surface of any solid which can be wetted by water. Insoluble solid particles pick up a small amount of moisture by physical adsorption. Soluble particles, as shown by ORR *et al.* (1958) with particles of $NaCl$, KCl, $(NH_4)_2 SO_4$, etc., remain solids at low humidities and become dissolved droplets at high humidities by undergoing a transition at some intermediate humidity. In the case of the heterogeneous catalysis of SO_2 in the atmosphere, previous investigators have shown that the rates of SO_2 oxidation were highest at humidities near saturation, but oxidation could not be detected when the experimental relative humidity was lower than about 70 per cent. Results from this investigation indicated, however, that SO_2 oxidation may proceed at all humidities, but the rate was much slower at low humidities than at saturation. FIGURE 12 shows the oxidation of SO_2 catalyzed by 1.2 mg of $NaCl$ aerosols at various relative humidities. Considering that sodium chloride particles undergo transition from solid crystals to solution drops at about 75 per cent relative humidity (ORR *et al.*, 1958), it is obvious that the curve at 81 per cent relative humidity looks drastically different from all other curves. But at 60 per cent relative humidity the oxidation proceeded at a measurable rate. Sodium

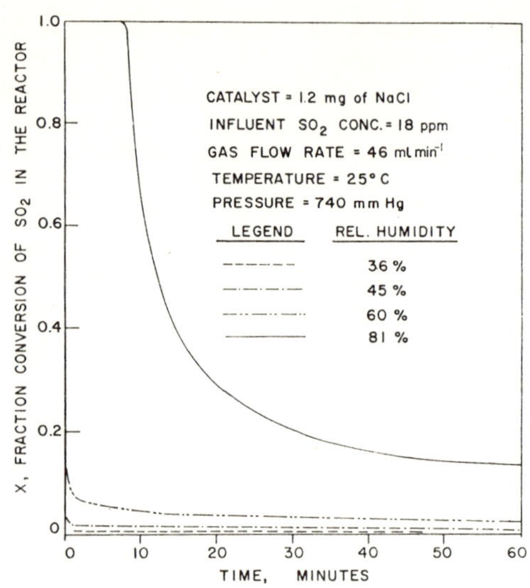

Fig. 12. Effect of relative humidity on the conversion of SO_2 in a reactor using NaCl aerosol in catalyst.

Table 2. Sulfate analysis of reactor contents for runs made with $CuCl_2$ and $MnCl_2$ aerosols*

Aerosol	Amount in Reactor (mg)	Total SO_2 removed (μg)	Total SO_4^{2-} formed (μg)	SO_2 Converted to SO_4^{2-} (%)†
$CuCl_2$	0.51	114	135	78.5
$CuCl_2$	0.68	150	195	87
$CuCl_2$	0.34	70	103	89
$MnCl_2$	0.40	90	128	95
$MnCl_2$	0.40	95	130	91

* Experimental conditions:
  Influent SO_2 concentration $= 14.4$ ppm
  Gas flow rate $= 46$ ml min^{-1}
  Temperature $= 23°C$
  Relative humidity $= 95$ per cent
  Pressure $= 740$ mm Hg
  Aerosol concentrations on beads $= 0.0425$ mg of $CuCl_2\,g^{-1}$ of beads and 0.040 mg of $MnCl_2\ g^{-1}$ of beads
  Duration of Runs $= 3$ h for the runs made with $MnCl_2$ aerosol; and 2 1/2 to 3 1/2 h for the runs made with $CuCl_2$ aerosol
† Percentage conversions were computed based on the maximum SO_4^{2-} that can be formed if all the SO_2 removed was converted to SO_4^{2-}. For example, for the run made with 0.51 mg of $CuCl_2$, the maximum SO_4^{2-} formation was $114 \times (96/64) = 171\ \mu$g. The actual SO_4^{2-} formation was 135 μg. The per cent conversion was therefore:

$$\frac{134}{171} \times 100 = 78.5 \text{ per cent}$$

chloride particles cannot transform into solution drops at this humidity; therefore, it is likely that oxidation of SO_2 occurred in the water film surrounding the particle surface. The formation of sulfuric acid in the water film could reduce the vapor pressure of the film surface slightly below that of the surrounding atmosphere, thus inducing more water vapor to diffuse towards, and condense upon the sodium chloride particle. At 25°C and 30 per cent relative humidity, ORR *et al.* (1958) estimated that a sodium chloride surface would be covered with a film of water averaging 1.42 molecules thick. The amount of dissolved SO_2 in this thin film must be extremely small. This scarcity of water might also explain why SO_2 oxidation at 36 and 45 per cent relative humidity only proceeded at a marginally detectable level. When the relative humidity was at 0 and 20 per cent, the SO_2 break-through curves were the same for the NaCl packed reactor and the dummy reactor, indicating that heterogeneous catalysis of SO_2 must be occurring within an aqueous catalyst solution. In the completely dry state the reaction

$$SO_2 + \tfrac{1}{2} O_2 \xrightarrow{\ \text{catalyst}\ } SO_3 \tag{11}$$

simply cannot proceed. The presence of even minute amounts of water, however, changed the reaction mechanism to

$$H_2SO_3 + \tfrac{1}{2} O_2 \xrightarrow{\ \text{catalyst}\ } H_2SO_4 \tag{12}$$

and markedly increased the rate of SO_2 oxidation.

4. EXTRAPOLATION OF RESULTS TO THE ATMOSPHERIC ENVIRONMENT

The reactor system and methods utilized were designed to reduce turbulent gas diffusion in the reactor. The highest gas flow rate was 147 ml min^{-1} which yielded an average interstitial gas velocity within the reactor of 40 cm min^{-1}. This velocity is approximately equivalent to the terminal settling velocity in air of a 15 μm spherical particle of unit density. A representative fog droplet of 10–20 μm diameter would settle in air with a terminal velocity of approximately 40 cm min^{-1}. Although a flow reactor was used in these experiments, while the atmospheric environment more closely resembles a huge batch reactor, the relative velocity between the aerosol phase and the gas phase was of the same order of magnitude in both situations. Furthermore, it was shown that the over-all rate of SO_2 oxidation is not limited by the gas phase diffusional resistance. Therefore, even if there were fluid dynamic differences between the aerosol–gas system in the atmosphere and in the laboratory, those differences would not affect the over-all rate of SO_2 oxidation. Therefore, we conclude that laboratory findings of this study can be extrapolated to the atmospheric environment.

With full hydration a deliquescent particle will grow to about five times its original size. For a $MnSO_4$ particle of 3 μm dia. and specific gravity of 3.0, the final diameter of the solution drop will be 15 μm, equivalent to a 125-fold increase in particle volume. The concentration of $MnSO_4$ in this solution drop is 24,000 ppm, which greatly exceeds the combined metal ion concentrations in a natural fog droplet. COUGHANOWR and KRAUS (1965) studied the effects of catalyst concentrations on the rate of oxidation of SO_2 in aqueous $MnSO_4$ solutions. They reported that from 0 to 100 ppm of $MnSO_4$

the reaction rate constant, k_0, was proportional to the square of the MnSO$_4$ concentration; above 100 ppm, k_0 increased less rapidly up to about 500 ppm after which it increased very slowly with catalyst concentration. These results suggest that catalyst concentrations were relatively high in solution drops used in our studies.

In order to estimate the rate of conversion of SO$_2$ to sulfuric acid in the atmosphere, the following adjustment was made. Let each solution drop in the experimental reactor retain its present full grown size, but reduce the weight of the original MnSO$_4$ particle about 50 times. In this way, the final MnSO$_4$ concentration in the drop is adjusted from the excessive 24,000 ppm to the adequate 500 ppm. In this way, the adjusted reaction rate equation can be obtained by multiplying the reaction rate constant of Equation (8) by 50. Therefore,

$$-R\left(\frac{\mu\text{g of SO}_2}{(\text{min})\,(\text{mg of MnSO}_4)}\right)$$

$$= 0.67 \times 10^{-2}\left(\frac{\text{m}^3 \text{ of air}}{(\text{min})\,(\text{mg of MnSO}_4)}\right) C\left(\frac{\mu\text{g of SO}_2}{\text{m}^3 \text{ of air}}\right) \tag{13}$$

The following assumptions were made for the conditions existing in a natural fog in an industrial atmosphere.

(1) Concentration of SO$_2$ in the air is 0.1 ppm.

(2) Average diameter of fog droplets is 15 μm.

(3) Half of the fog droplets contain catalyst capable of oxidizing SO$_2$ to H$_2$SO$_4$. The catalyst concentration within these fog droplets is equivalent to 500 ppm of MnSO$_4$. This droplet could be nucleated by a single particle of MnSO$_4$ having a diameter of 0.55 μm and a specific gravity of 3.0, or it could contain other types of catalyst with strength equivalent to 500 ppm of MnSO$_4$.

(4) The fog concentration is 0.2 g of water per cubic meter of air.

With these assumptions the catalyst concentration is calculated to be equivalent to 50 μg of MnSO$_4$ per cubic meter of ambient air. The NATIONAL AIR SAMPLING NETWORK (1962, 1966) indicates that the maximum concentration for manganese in urban air is 9.98 μg m^{-3}. The maximum concentrations for other metals that could act as catalysts for the oxidation of sulfur dioxide are: copper (10 μg m^{-3}), zinc (58 μg m^{-3}), iron (74 μg m^{-3}), and lead (17 μg m^{-3}). In addition, organic suspended particulate matter could also oxidize SO$_2$ to sulfuric acid. Therefore, 50 μg of MnSO$_4$ per cubic meter of air as the equivalent catalyst concentration in a heavily polluted industrial atmosphere is not excessive. With these assumptions and equation (13), the rate of conversion of SO$_2$ to sulfuric acid was calculated to be 2 per cent h^{-1}.

JOHNSTONE and COUGHANOWR (1958) estimated the rate of atmospheric catalytic oxidation of SO$_2$ under saturation conditions to be around 1 per cent per min when catalyst concentration was equivalent to 50 μg m^{-3} of MnSO$_4$ and SO$_2$ concentration was 1 ppm. On the other hand, MATTESON et al. (1969) estimated the same conversion rate to be around 0.5 per cent per day under similar environmental conditions. Thus, the same rate estimated by one group of investigators was three thousand times larger than that estimated by another group. The work reported by MATTESON et al. (1969) appeared to have suffered from inferior data collecting techniques, while Johnstone and Coughanowr's work was based on the doubtful assumption that the reaction rate was controlled by liquid phase gas diffusion.

GARTRELL *et al.* (1963) studied the oxidation of SO_2 in coal-burning power plant plumes. The highest SO_2 oxidation rate, 55.5 per cent in 108 min, was observed during foggy meteorological conditions. The concentration of particulate matter in power plant plumes is obviously much higher than that in ambient air. Assuming catalyst concentration in the plume is equivalent to 500 μg of $MnSO_4$ per cubic meter of plume air, the rate of conversion calculated from equation (13) is 20 per cent h^{-1}, which is close to the approximately 30 per cent h^{-1} oxidation rate reported by Gartrell *et al.*

CONCLUSIONS

A laboratory investigation of the heterogeneous catalysis of SO_2 in the atmosphere was conducted. We conclude the following.

(1) Aerosols of $MnSO_4$, $MnCl_2$, $CuSO_4$ and $NaCl$ acted as true catalysts when catalytic oxidation of SO_2 occurred within their aqueous solutions. Among the four salts tested, $MnSO_4$ was found to be the most effective catalyst and $NaCl$ the least effective.

(2) The over-all rate of absorption and oxidation of SO_2 by aqueous catalyst solution drops was limited by the chemical reaction itself.

(3) When $MnSO_4$ aerosol was used as a catalyst, the over-all reaction rate was found to be first order with respect to SO_2 concentration in the gas phase.

(4) Relative humidity of the air appeared to exert the strongest influence on the rate of SO_2 oxidation in air. Higher rates of oxidation always accompanied higher relative humidities at an air temperature of 23°C and ambient air pressure of 740 mm Hg.

(5) Cupric chloride ($CuCl_2$) aerosol acted as a primary reactant with SO_2. SO_2 was oxidized to sulfuric acid while $CuCl_2$ was reduced to $CuCl$.

(6) Laboratory results were extrapolated to the urban industrial atmospheric environment to estimate the rate of absorption and oxidation of SO_2 by fog droplets in a natural fog. This rate was estimated to be 2 per cent h^{-1}.

REFERENCES

AMDUR M. O. (1969) Toxicologic appraisal of particulate matter, oxides of sulfur, and sulfuric acid. *J. Air Pollut. Control Ass.* **19,** 638–644.

AMDUR M. O. and UNDERHILL D. (1968) The effect of various aerosols on the response of guinea pigs to sulfur dioxide. *Archs Environ. Hlth* **16,** 460–468.

ANDERSON L. B. and JOHNSTONE H. F. (1955) Gas absorption and oxidation in dispersed media. *Am. Inst. Chem. Engng J.* **1,** 135–141.

BASSETT H. and HENRY J. (1935) Formation of dithionate by the oxidation of sulfurous acid and sulphites. *J. Chem. Soc.* 914–929.

BASSETT H. and PARKER W. G. (1951) Oxidation of sulfurous acid. *J. chem. Soc.* 1540–1560.

CHENG R. T., FROHLIGER J. O. and CORN M. (1971) Aerosol stabilization for laboratory studies of aerosol–gas interactions. *J. Air Pollut. Control Ass.* **21,** 138–142.

COUGHANOWR D. R. (1956) Oxidation of sulfur dioxide in drops. Ph.D. Thesis, University of Illinois.

COUGHANOWR D. R. and KRAUS F. E. (1965) The reaction of SO_2 and O_2 in aqueous solution of $MnSO_4$. *Ind. Engng Chem. Fundam.* **4,** 61–66.

GARTRELL F. E., THOMAS F. W. and CARPENTER S. B. (1963) Atmospheric oxidation of SO_2 in coal burning power plant plumes. *Am. ind. Hyg. Ass. J.* **24,** 113–120.

GERHARD E. R. and JOHNSTONE H. F. (1955) Photochemical oxidation of sulfur dioxide in air. *Ind. Engng Chem.* **47,** 972–976.

JOHNSTONE H. F. and COUGHANOWR D. R. (1958) Absorption of sulfur dioxide from air. *Ind. Engng Chem.* **50,** 1169–1172.

JOHNSTONE H. F. and LEPPLA P. O. (1934) The solubility of sulfur dioxide at low partial pressure. The ionization control and heat of ionization of sulfurous acid. *J. Am. chem. Soc.* **56,** 2233.

JOHNSTONE H. F. and MOLL A. J. (1960) Formation of sulfuric acid in fogs. *Ind. Engng Chem.* **52,** 861–863.

JUNGE C. E. (1960) Sulfur in the atmosphere. *J. geophys. Res.* **65,** 227–237.

JUNGE C. E. and RYAN T. G. (1958) Study of the SO_2 oxidation in solution and its role in atmospheric chemistry. *Q. J. R. Met. Soc.* **84,** 46–55.

LEVENSPIEL O. (1962) *Chemical Reaction Engineering* pp. 449–476. J. Wiley, New York.

MATTESON M. J., STOBER, W. and LUTHER H. (1969) Kinetics of the oxidation of sulfur dioxide by aerosols of manganese sulfate. *Ind. Engng Chem. Fundam.* **8,** 677–687.

O'KEEFFE A. E. and ORTMAN G. C. (1966) Primary standards for trace gas analysis. *Anal. Chem.* **38,** 760–763.

ORR C. JR., HURD F. K. and CORBETT W. J. (1958) Aerosol size and relative humidity. *J. Colloid Sci.* **13,** 472.

PRITCHETT P. W. (1961) The catalyzed reaction of oxygen with sulfurous acid and its effect on the absorption of oxygen and sulfur dioxide into water. Ph.D.Thesis, University of Delaware.

SCARINGELLI F. P., FREY S. A. and SALTZMAN B. E. (1967) Evaluation of Teflon permeation tubes for use with sulfur dioxide. *Am. ind. Hyg. Ass. J.* **28,** 260–266.

SCHWARTZ C. E. and SMITH J. M. (1953) Flow distribution in packed beds. *Ind. Engng Chem.* **45,** 1209–1218.

TICHACEK, L. J. (1963) Selectivity in experimental reactors. *Am. Inst. chem. Engng J.* **9,** 394–399.

U. S. DEPT. OF HEALTH, EDUCATION, AND WELFARE (1962) *Air Pollution Measurements of National Air Sampling Network*, 1957–1961, Div. of Air Pollution, Cincinnati, Ohio.

U. S. DEPT. OF HEALTH, EDUCATION, AND WELFARE (1965) *Selected Methods for the Measurement of Air Pollutants*. Environmental Health Series, Air Pollution 999–AP–111 (1965).

U. S. DEPT. OF HEALTH, EDUCATION, AND WELFARE (1966) *Air Quality Data* 1964–1965. Div of Air Pollution, Cincinnati, Ohio.

VAN DER LAAN, E. TH. (1958) Notes on the diffusion-type model for the longitudinal mixing in flow. *Chem. Engng Sci.* **7,** 187–191.

The Effect of Water Vapor on Ozone Synthesis in the Photo-oxidation of Alpha-pinene

Lyman A. Ripperton and Daniel Lillian

In pioneering work on the elucidation of photochemical smog formation Haagen-Smit and Bradley[1] found that in the systems studied water vapor was essential for the synthesis of ozone (O_3). In most studies prior to 1963 dry or humid air was used somewhat arbitrarily and the effect of varying the humidity was not reported.[2]

Ripperton and Jacumin[3] utilized the system, hexane-1 + NO_2 + $h\nu$, at concentrations of 2 ppm and 1 ppm respectively and varied water vapor concentration and temperature. Water vapor at 0.0000, 0.0045, 0.0090, 0.0135, and 0.0180 grams per gram of dry air and temperatures of 14°, 25°, and 35°C were utilized. At all tempera-tures a maximum concentration of O_3 was achieved at the level of 0.0090 g of water/g of dry air.

Dimitriades[4] reported substantial effects of water vapor on systems of 1.65 ppm ethylene and 0.50 ppm nitric oxide (NO). Increasing relative humidities from 1.5 to 35% (93°F) increased the rates of oxidant and form-aldehyde formation as well as the rate of NO photo-oxidation.

Bufalini and Altshuller[5] observed no effect of water vapor on the rate of oxidation of NO or on the photo-oxidation rate of olefin in systems of 2 ppm NO and various concentrations of either 2,3-dimethyl-2-butene or 1-butene or n-butane.

Wilson and Levy[6] and Wilson[7] reported significant effects of water vapor on systems of 4 ppm 1-butene and 1 ppm NO. Increasing relative humidity from 0.65% (82°F) was found to decrease (1) the maximum and steady-state oxidant concentrations, (2) the half/life of the NO, (3) the time for utilization of half that amount of 1-butene utilized in reaching steady-state 1-butene concentrations, and (4) the time required for oxidant and nitrogen dioxide (NO_2) concentrations to reach their maximum values.

The study reported here was undertaken to examine the effects of water vapor on the synthesis of O_3 in the system NO_2 + alpha-pinene + $h\nu$.

Procedure

Five $3 \times 2 \times 2$ complete factorial experiments were performed in duplicate to obtain concentration-time profiles of selected reactants and products. Water vapor concentrations of 0.000, 0.0090 and 0.0180 g/g of dry air were utilized. Alpha-pinene at zero and 50 parts per hundred million (pphm) and NO_2 at zero and 10 pphm were utilized. Due to restrictions placed by the equipment utilized not all measurements could be made simultaneously. Ten runs were made for each of the 12 reaction systems. Two runs for each set of reactant concentrations were made to measure (1) oxidants alone, (2) condensation nuclei and oxidants, (3) O_3 and oxidants, (4) NO_2 and oxidants, (5) alpha-pinene, NO, and oxidants.

Reactions were carried out in 150-liter 5-mil FEP type C Teflon bags. Temperature was maintained at 21° ± 1°C in the irradiation chamber; in the Teflon bags temperature was 25° ± 2°C. Irradiation was provided by 40-watt cool-white fluorescent lamps and 2 E-H-1 mercury vapor lamps. The ϕk_a for NO_2 photolysis was determined experimentally to be 2.8 hr^{-1}, representing an intensity of approxi-

mately 15% of noonday sunlight.[2]

Gas blends were made in Matheson zero air cleaned with anhydrous calcium sulfate, activated charcoal and 4A molecular sieve. Alpha-pinene was introduced with a 1-microliter syringe, NO_2 with a permeation tube[8] and water was evaporated into the air in an in-line flask on a hot plate.

Oxidant was measured with the Mast (coulometric) O_3 meter. Condensation nuclei were measured with a General Electric type CN condensation nuclei detector. Ozone was measured with a chemiluminescent O_3 detector. Nitrogen dioxide and NO were measured with a modified Saltzman reagent in a continuous analyzing instrument. Alpha-pinene was measured with a gas chromatograph equipped with a flame ionization detector.

The data were analyzed statistically using Tukey's method of multiple comparison[9] between the three different humidity levels. Comparisons were made of the mean values of a given variable obtained with a given set of experimental conditions. The word "significant" in the following discussion refers to the 0.05 confidence level for all cases except those noted as 0.01 in the text.

Results and Discussion

Oxidants in Systems of 10 pphm NO_2

In all experiments, systems of 10 pphm NO_2 formed oxidants at each of the three humidity levels studied. The average concentration-time profiles obtained for these systems are presented in Figure 1. The oxidant concentration increased to its near-maximum value within the first few minutes of irradiation and remained nearly constant thereafter. This attainment of steady-state is consistent with accepted theory:

$$NO_2 + h\nu = NO + O \qquad (1)$$

$$O + O_2 + M = O_3 + M \qquad (2)$$

$$O_3 + NO = NO_2 + O_2 \qquad (3)$$

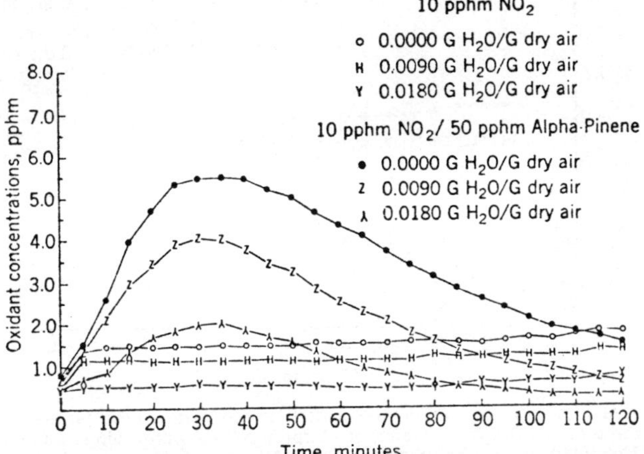

Figure 1. Oxidant concentration-time profiles at three humidities for systems of 10 pphm NO₂ and 10 pphm NO₂/50 pphm alpha-pinene. The mean curves of duplicate runs are drawn.

The highest oxidant concentrations were observed for the dry systems throughout the entire irradiation period. Water's statistically significant suppression of the overall concentration-time profiles of oxidants to lower concentrations, without alteration of their steady-state shape, suggests that water influences the primary process (Eq. 1) directly or indirectly. This follows since participation of water vapor in reactions (2) or (3) is either theoretically or experimentally contra-indicated. The effect could be caused by reduction of the quantum yield of atomic oxygen from NO₂ photolysis or reduction of the effective concentration of NO₂.

The only reasonable reactions involving water and the reactants of Eq. (2) are participation of water vapor as the third body (M) or direct reaction of water with atomic oxygen to form hydroxyl radicals. Participation of water as the third body (M) would be expected to enhance ozone formation.[2] Since this is contrary to the observed results, such participation is unlikely.

A reaction between water and the (3P) atomic oxygen generated from the photolysis of NO₂ has been theoretically and experimentally shown to be insignificantly slow at ambient temperatures.[2]

Although photolysis of NO₂ is the primary photochemical reaction most cited in air pollution literature, there is evidence that collisional deactivation is an important competing primary process. This is seen in a comparison of the quantum yields for dissociation of NO₂ in nitrogen at 4047 and 3660 Å. These quantum yields were reported by Leighton[2] as 0.36 and 0.96, respectively. At wavelengths greater than 3600 Å processes such as resonance transfer, fluorescence, and collisional deactivation apparently become more important and at 3660 Å are very important. Because of its polarity, the water molecule should serve as a more efficient body for collisional deactivation than molecules of nitrogen and oxygen.[10]

A reasonable mechanism whereby water affects removal of NO₂ from availability for photolysis at a rate

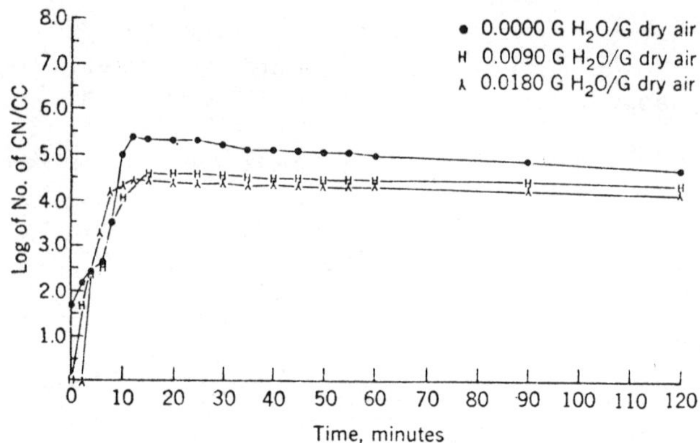

Figure 2. Condensation nuclei (CN) concentration-time profiles at three humidities for systems of 10 pphm NO_2/50 pphm alpha-pinene. The mean curves of duplicate runs are drawn.

which allows a near steady-state oxidant concentration would be one involving one or more equilibria which maintain the NO_2 concentration below its corresponding concentration in a dry system. A series of reactions which could accomplish this is:

$$NO_2 + h\nu = NO + O \qquad (1)$$

$$O + O_2 + M = O_3 + M \qquad (2)$$

$$O_3 + NO = NO_2 + O_2 \qquad (3)$$

$$NO + NO_2 = N_2O_3 \qquad (4)$$

$$N_2O_3 + H_2O = 2HNO_2 \qquad (5)$$

$$NO_2 + O_3 = NO_3 + O_2 \qquad (6)$$

$$NO_3 + NO_2 = N_2O_5 \qquad (7)$$

$$N_2O_5 + H_2O = 2HNO_3 \qquad (8)$$

With increasing water vapor concentration, the equilibria of reactions (5) and (8) would be shifted to the right and the amount of free NO_2 available for photolysis would be accordingly reduced. Published equilibrium constants for reactions (5) and (7) are 1.65 atm^{-1} (based on the reaction $NO + NO_2 + H_2O = 2HNO_2$) and 2×10^6, respectively.[2] Both constants in-

dicate the feasibility of their associated reactions as mechanisms for decreasing NO_2 concentrations with water vapor. Since the latter constant is much higher than the value which would be consistent with field data, Leighton suggests that equilibrium is never reached in reaction (8).

Oxidants in Systems of 10 pphm NO_2/50 pphm Alpha-Pinene

In all experiments, systems of 10 pphm NO_2/50 pphm alpha-pinene formed oxidants at each of the three humidity levels studied. The concentration-time profiles obtained for these systems are also shown in Figure 1. At each humidity, the oxidant rose to a maximum value 30–40 min into the run. After reaching this value, it decreased slowly for the remainder of the irradiation time. Net oxidant production was much greater for these systems than for the NO_2 systems—at least in the first 60 min of irradiation.

In chemical models of smog such as the system used in this study, the oxidant concentration-time profiles may be explained by free radical reactions which accompany the photo-oxidation of the olefin. The initial formation of free radicals, $R\cdot$, can be attributed to a reaction between alpha-pinene and the atomic oxygen generated from NO_2 photolysis:

$$O + \text{alpha-pinene} = R\cdot \qquad (9)*$$

Formation of ozone in excess of that produced following NO_2 photolysis can occur by reactions between free radicals and molecular oxygen:

$$R\cdot + O_2 = ROO\cdot \qquad (10)$$

$$ROO\cdot + O_2 = O_3 + RO\cdot \qquad (11)$$

There is criticism of reaction (11) on the grounds that it must be endothermic but it is still evoked in various "smog" mechanisms.[11] Leighton[2] said there is good experimental evidence supporting its occurrence at ambient temperatures.

In the oxidant curves shown in Figure 1, although neither the time for maximum oxidant concentrations nor the general shape of the curves is affected appreciably by water vapor, increasing water levels suppressed the overall profiles to lower oxidant concentrations. This effect was significant at the .01 confidence level and in the early stages of irradiation may have been due largely to the effect of water on NO_2. Additional water participation is indicated in one or more processes subsequent to the inorganic reactions.

It is a well known phenomenon that condensation of water vapor occurs on photochemical aerosol at partial pressures of water vapor less than saturation.[12] Water vapor will condense with oxygenated intermediates in the present study. From a consideration of surface tension, one would expect the formation of droplets with water at the center and organic compounds on the surface.[12] The polar functional groups of the organic compounds should be oriented toward the aqueous center of the droplets. A suppression of oxidant vapor pressure may then be explained by physical removal of reactive intermediates from the gas phase. The high dielectric constant of water may induce competing polar reactions[13] which may not lead to the formation of O_3. Furthermore, the reactive sites of the organic molecules would be protected from interaction with the gas phase.

Oxidants in Systems of Zero Air and Systems of 50 pphm Alpha-pinene

In all experiments, at all humidities, no system of zero air nor of 50 pphm alpha-pinene exhibited oxidant concentrations greater than 0.1 pphm (Full scale deflection of the instrument was 10 pphm) within the first 70 min of irradiation. Several runs exhibited a gradual increase of oxidants concentration throughout the remainder of the irradiation which is thought to be a manifestation of slight contamination.

Condensation Nuclei in All Systems

All systems studied formed condensation nuclei (CN). Except in the NO_2/alpha-pinene systems the formation of CN is not readily explained although similar behavior has been observed by others. Goetz and Klejnot[14] observed aerosol formation upon irradiation of olefin systems in the absence of NO_2. Decker and Page[15] have observed the rapid formation of CN upon irradiation of NO_2 systems in the absence of olefin.

Consistent with theory [16–18] systems of 10 pphm NO_2/50 pphm alpha-pinene exhibited the highest net production of CN at each humidity. This was especially noticeable in the first few minutes of irradiation when contamination effects were probably minimal. The concentration-time profiles for

---

*By analogy with Cvetanovic's proposed mechanism for the reaction of atomic oxygen with various olefins[2] one would expect a ring cleavage adjacent to the double bond and the resulting free radical to have a diradical configuration—·OR.

this system are shown in Figure 2.

By analogy to other described-photochemical smog systems,[2] aerosol formation by NO_2/alpha-pinene systems may be explained in part by a peroxy-free radical polymerization:

$$ROO\cdot + pinene = ROO\text{-}pinene \qquad (12)$$

$$ROO\text{-}pinene + O_2 =$$
$$ROO\text{-}pinene\text{-}OO\cdot \qquad (13)$$

$$ROO\text{-}pinene\text{-}OO\cdot + pinene =$$
$$ROO\text{-}pinene\text{-}OO\text{-}pinene, \text{ etc.} \qquad (14)$$

A reaction between O_3 and alpha-pinene has been shown to lead to the formation of light scattering aerosol[16,17] and may be significant also.

Went's observation that O_3 and alpha-pinene react to form a slight aerosol was corroborated in this study. When a blend of 1 ppm NO_2/1 ppm alpha-pinene was irradiated in a 50-l flask, a slight bluish haze become visible in a Tyndall beam within 15 min. In a number of experiments, Tyndall beams were observed after the reaction of alpha-pinene with O_3 utilizing-concentrations down to 10 pphm of each.

In the NO_2/alpha-pinene systems, increasing humidity suppressed net CN formation. This may be accounted for by the NO_2-water interaction previously discussed which effectively suppresses atomic oxygen formation. Alternatively, the reduction of O_3 concentrations by inhibition of the free-radical mechanism can explain this behavior. Since the CN counter measures number and not mass concentration, the effect may have been due to water's promoting coalescence also.

Ozone in Systems of 10 pphm NO_2

At all humidities, systems of 10 pphm NO_2 exhibited the characteristic build up and maintenance of a steady state O_3 concentration consistent with reactions (1) through (4). Figure 3 shows the mean O_3 concentration-time profiles for these systems. From this figure a water effect is apparent. Increasing levels of water suppressed O_3 concentrations. The effect of increasing levels of water vapor was similar in trend to the effect observed for oxidants. As such, an effective shift to the left of the equilibrium — $NO_2 + O_2 = NO + O_3$ — is again in-

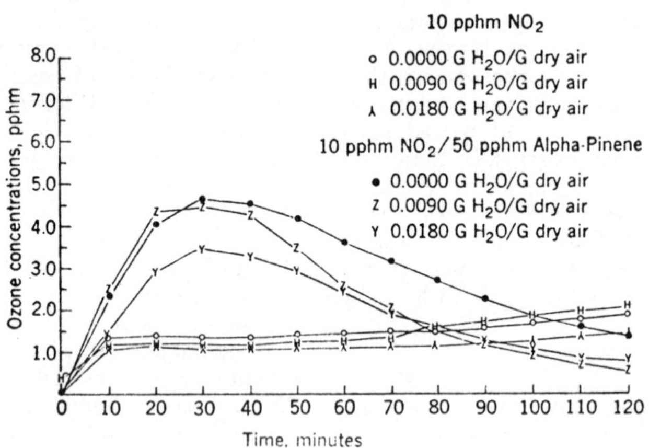

Figure 3. Ozone concentration-time profiles at three humidities for systems of 10 pphm NO_2 and 10 pphm NO_2/50 pphm alpha-pinene. The mean curves of duplicate runs are drawn.

dicated. As previously discussed, this shift may be accomplished by a reduction of the primary quantum yield and/or removal of NO_2 from photolysis by reaction(s) involving water.

Ozone in Systems of 10 pphm NO_2/50 pphm Alpha-pinene

The mean O_3 concentration-time profiles for systems of 10 pphm NO_2/50 pphm alpha-pinene at three humidities are shown in Figure 3. As in the oxidant experiments, at each humidity, the O_3 concentration is seen to build up to a maximum value between 30 to 40 min after the start of irradiation and then decrease thereafter. Comparison of these profiles with those of the 10 pphm NO_2 systems reveals the same relationship which exists between the oxidant profiles of the NO_2 and NO_2/alpha-pinene systems: within the first 70 min of irradiation, the O_3 production of the systems containing alpha-pinene was significantly greater. As in the oxidant experiments, this behavior is consistent with a free-radical mechanism.

The effects of water vapor on these systems were statistically significant. As the corresponding data for the oxidant experiments indicate, these data indicate that water may participate in either the inorganic or organic set of reactions, or both.

Ozone in Systems of Zero Air and Systems of 50 pphm Alpha-pinene

Of the 12 runs constituting the total number for the zero air systems and alpha-pinene systems combined, nine of them did not exhibit O_3 concentrations greater than 0.1 pphm within the first hour of irradiation; three of them had attained O_3 concentrations of 0.2 pphm. These low readings within the first irradiation and the gradually increasing O_3 concentration exhibited by a few runs after this time indicate a minor degree of contamination.

Comparison Between Oxidant and Ozone Profiles for Systems of 10 pphm NO_2

At the zero level of water, the oxidant profiles are higher than the O_3 profiles. At the intermediate level of water, the oxidant profiles are still higher than the O_3 profiles, but not as much. At the highest humidity level, the oxidant profiles are slightly higher than the O_3 profiles. Since the Mast instrument responds, to varying degrees, to all oxidants—it is approximately $\frac{1}{10}$ as sensitive to NO_2 as to O_3, for example[18]—the higher readings of the Mast instrument than of the Regener device are expected. The decrease in oxidant/O_3 ratio with increasing water vapor pressure warrants further discussion.

The ratios of the mean/time oxidant concentrations to mean/time O_3 concentrations were $2.4/1.4 = 1.7$ and $1.3/1.1 = 1.2$ at the lowest and highest humidity levels respectively. The mean/time NO_2 readings for the lowest and highest humidity levels were 7.7 and 8.7 pphm, respectively. A comparison of these values with the mean/time oxidant and O_3 values indicates that at the lowest humidity, the total oxidant concentration can be accounted for by the O_3 and NO_2 present. At the highest humidity the O_3 alone can account for 1.1 of 1.3 pphm or 85% of the total oxidant concentration. Since a mean/time value of 8.7 pphm of NO_2 was obtained at the highest humidity and this did not have as large an effect as expected on the readings of the Mast instrument but did diazotize the NO_2 reagent, one would suspect the presence of nitrous acid. Nitrous acid reacts with the diazotizing reagent (positive NO_2 reaction) but would not be detected by the Mast oxidant meter.† This

---

† Although the analysis of HNO_2 can be accomplished by iodometric means experience with NO_2 has shown that nitrite in the Mast instrument does not produce iodine on a mole for mole ratio.

is evidence supporting the proposed mechanism for oxidant and O_3 depression by water vapor in the inorganic gas blends.

Comparison Between Oxidant and Ozone Profiles For Systems of 10 pphm NO_2/50 pphm Alpha-pinene

At each humidity, oxidant readings are higher than corresponding O_3 readings, but with increasing humidity, their difference becomes less. The average mean/time oxidant, O_3, and NO_2 readings for the low- and high-humidity levels and the oxidant readings attributable to NO_2 (labeled O_xNO_2) and to NO_2 plus O_3 are presented in Table I. Comparisons of the data in this table indicate that the O_3 and the NO_2 oxidant equivalent cannot account for the total oxidant reading at the low humidity. The excess oxidant is, by analogy with the results of other studies, considered to be organic peroxy compounds. At the high humidity, the O_3 alone accounts for all oxidant measured. This indicates the absence of any amount of NO_2 or organic peroxy compounds detectable by the Mast instrument. In the presence of organic vapors, a number of nitroxy compounds capable of diazotizing the NO_2 reagent are formed.[2] The presence of these substances and nitrous acid could account for a positive NO_2 reaction without an accompanying effect on the oxidant measurement. The fact that no detectable concentration of peroxy compounds occurred indicates involvement of water in the organic reactions subsequent to the olefin-atomic oxygen reaction.

Nitrogen Dioxide in Systems of 10 pphm NO_2

In systems of 10 pphm NO_2, NO_2 concentrations were found to decrease rapidly at each humidity during the first 10 minutes of irradiation. After this time concentrations continued to decrease, but at a much slower rate, indicating an approach to a steady state.

The concentration-time profiles for these systems are shown in Figure 4. A small water effect is apparent. Increasing levels of water vapor enhanced NO_2 concentrations throughout the irradiation time. This effect suggests again the two following mechanistic possibilities: First, the presence of water decreases the primary quantum yield of NO_2 allowing higher concentrations of NO_2 to be maintained.. Second, NO_2 is removed from photolysis by a complex reaction with water to form products which effect a positive Saltzman reagent response. Materials such as nitrous acid and any species capable of forming the nitrosonium ion in solution can effect this response. The arguments presented in the section on the comparison of oxidant and O_3 profiles would not rule out some activity of the first kind, but definitely support the second hypothesis.

Nitrogen Dioxide in Systems of 10 pphm NO_2/50 pphm Alpha-pinene

At each humidity, systems of 10 pphm NO_2/50 pphm alpha-pinene exhibited an approximate 90% destruction of NO_2 by the completion of the irradiation. Consistent with theory, a steady state was never approached, and NO_2

Table I

Humidity level	Oxidants (pphm)	O_2 (pphm)	NO_2 (pphm)	O_xNO_2 (pphm)	O_xNO_2 + O_2 (pphm)
Low	4.3	2.8	3.8	0.4	3.2
High	1.7	1.8	4.1	0.4	2.2

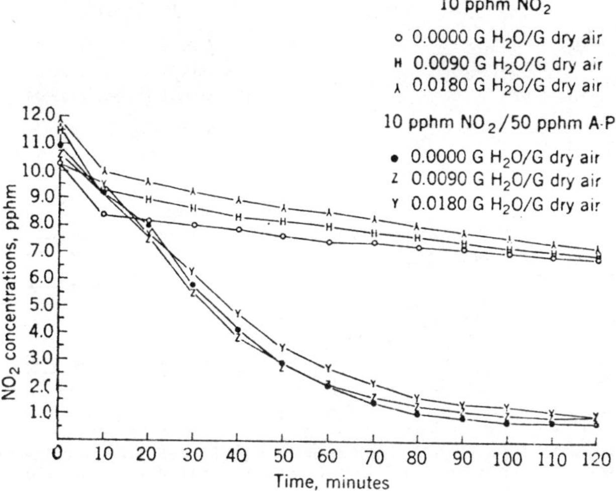

Figure 4. NO₂ concentration-time profiles at three humidities for systems of 10 pphm NO₂ and 10 pphm NO₂/50 pphm alpha-pinene. The mean curves of duplicate runs are drawn.

concentrations decreased continuously from the beginning of the irradiation.

Statistical analysis showed for the NO₂ systems that between the lowest and highest humidity levels, increasing water vapor decreased net NO₂ utilization and had no significant effect between adjacent levels. For systems of NO₂/pinene no effects of water were statistically significant.

Nitric Oxide in Systems of 10 pphm NO₂

Concentration-time profiles for systems of 10 pphm NO₂ at three humidities are shown in Figure 5. At each humidity the NO concentration is seen to build up its maximum value in the first 10 minutes of irradiation. After this time the concentrations remained nearly constant, consistent with the equilibrium:

$$NO_2 + O_2 = NO + O_3 \qquad (4)$$

The poor precision of the data for these systems, particularly at the intermediate level of water, precluded the observation of a water effect. The significant effects of water vapor on

oxidant, O₃ and NO₂ concentrations indicate that water vapor should have an effect on NO concentrations also. This follows from the dependence of NO on NO₂ and O₃ [reaction (4)] as well as from the involvement of NO in the proposed equilibria involving the oxides of nitrogen and water vapor.

The positive NO readings at time-zero are undoubtedly artifacts of the analytical method caused by the scrubbing column used to remove NO₂. Mueller, *et al.*,[19] have demonstrated that NO is synthesized in the Saltzman scrubbing reagent (e.g., $3NO_2 + H_2O = 2HNO_3 + NO$) and will be measured in a serial absorption system.

Nitric Oxide in Systems of 10 pphm NO₂/50 pphm Alpha-pinene

The NO concentration-time profiles for systems of 10 pphm NO₂/50 pphm alpha-pinene at three humidities are also shown in Figure 5. Consistent with theory, the NO concentrations built up rapidly within the first few minutes of irradiation and declined rapidly thereafter—probably to near

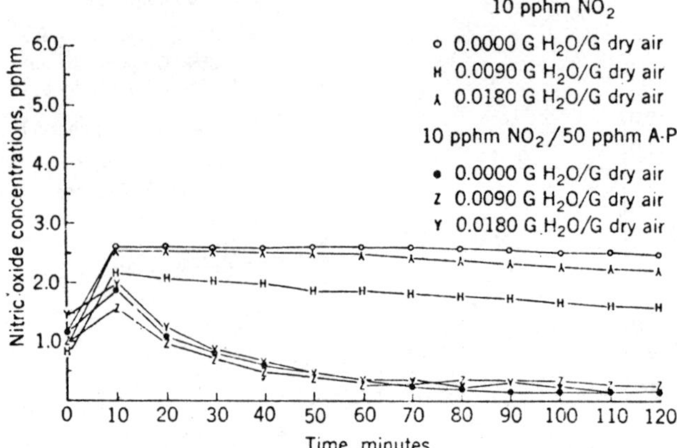

Figure 5. Nitric oxide concentration-time profiles at three humidities for systems of 10 pphm NO₂ and 10 pphm NO₂/50 pphm alpha-pinene. The mean curves of duplicate runs are shown.

zero considering the analytical artifact previously discussed. Again poor precision of data precluded the observation of a water effect if such was operative. Statistical analysis showed no NO changes as a function of humidity to be significant.

Alpha-pinene in All Systems

Alpha-pinene was not detected in irradiated systems of zero air and irradiated systems of 10 pphm NO₂ at all humidities. The chromatograph was capable of detecting concentrations as low as a few ppb.

No changes in alpha-pinene behavior with increasing humidity were statistically significant in any system.

Although the lack of effects for the pinene systems could be attributable to the poor precision between duplicate runs, there is no theoretical or experimental evidence supporting the photooxidation of alpha-pinene by radiation in the ground-level solar spectrum in the absence of a primary absorber such as NO₂.

Systems of 10 pphm NO₂/50 pphm alpha-pinene exhibited expected behavior as can be seen in Figure 6

Alpha-pinene concentrations are seen to decrease during irradiation at all humidities.

The reduction of net O_3 production effected by increasing water vapor pressure did not decrease the rate of alpha-pinene utilization. This was surprising since alpha-pinene reacts with O_3 ($k = 10^5$ 1 mole^{-1} sec^{-1}).[20] Wilson observed higher steady-state olefin concentrations with increasing humidity levels in a dynamic reaction system.[7]

Summary and Conclusions

Upon irradiation, systems containing NO₂ and alpha-pinene formed oxidants, O_3, condensation nuclei, and NO. Based on the differences between simultaneous oxidant and O_3 measurements, the formation of peroxide-like compounds may be inferred. During the course of the irradiation, NO₂ and alpha-pinene were consumed. The concentration-time profiles of all variables were characteristic of those exhibited by typical photochemical smog systems.

An effect of water vapor on the systems studied was demonstrated. Increasing humidity decreased net mean/time oxidant and O_3 production an'

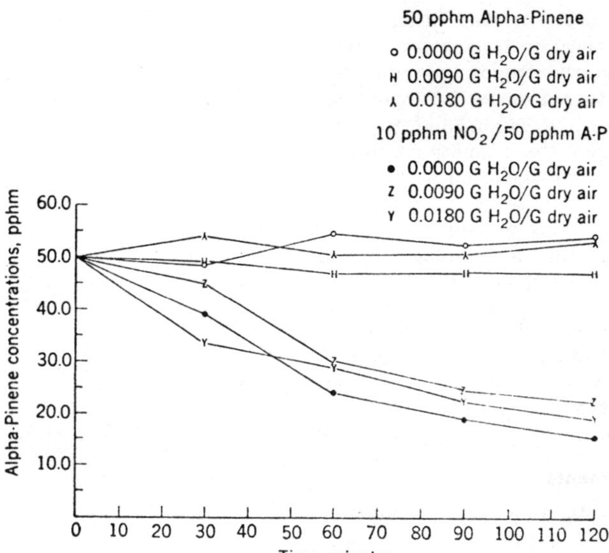

Figure 6. Alpha-pinene concentration-time profiles at three humidities for systems of 50 pphm alpha-pinene and 10 pphm NO₂/50 pphm alpha-pinene. The mean curves of duplicate runs are drawn.

net maximum condensation nuclei production. These effects were significant at a 0.05 confidence level. Effects of water vapor on average mean/time NO_2, NO and alpha-pinene concentrations were insigificant at this level. The oxidant to O_3 ratio was found to decrease with increasing humidity.

Based on significant decreases in net oxidant and O_3 production and NO_2 consumption with increasing water vapor concentrations in systems of NO_2 alone, strong support is given that water manifests an effect on pertinent inorganic reactions, although the experimental data indicate additional participation in the organic reactions.

A mechanism whereby water affects oxidant, O_3, and condensation nuclei concentrations, but not alpha-pinene concentrations is unlikely. Since O_3 reacts with alpha-pinene, a difference in O_3 concentration should produce a difference in alpha-pinene concentration. The large experimental error of the analytical method could account for the apparent lack of a significant water effect on the latter. It seems less likely that different mechanisms at the different humidity levels produced the same rate of alpha-pinene utilization.

The differences in results of this study from those of studies performed by others can be explained -currently only by difference in identity of initial reactants, concentrations utilized and possibly by the experimental conditions. The differences observed by the several workers cited in the literature review are undoubtedly real and more research must be performed first with simple systems and then with systems containing mixtures of organic vapors before the effect of water vapor on the photo-oxidation of organic compounds in ambient air can be predicted.

47

Since O_3, oxidants, and other products of photochemical smog are rarely emitted into the troposphere directly but are rather formed by reactions involving organic vapors and the oxides of nitrogen, the ultimate control of photochemical air pollution will depend largely on the emission standards set for the latter two variables. To the extent that the meteorology of a given area will influence the levels at which these standards are set, the significant effect of water vapor on the concentrations of photochemical smog substances would suggest that the ambient humidity should be considered also.

Acknowledgments

This investigation was conducted as a part of a graduate training program and was supported in part by Public Health Service Training Grant Number ES 61 from the Division of Environmental Health Sciences, National Institutes of Health. A substantial part of this work was carried out also under NSF Grant GA-1022: Chemical and Environmental Factors affecting Ozone Concentrations in the Lower Troposphere.

References

1. A. J. Haagen-Smit, and C. E. Bradley, "Ozone formation in photochemical oxidation of organic substances," *Ind. Eng. Chem.* **45**, 2986 (1953).
2. P. A. Leighton, *Photochemical Aspects of Air Pollution.* New York, Academic Press, Inc., 1961.
3. L. A. Ripperton, and W. J. Jacumin, "The Effect of Humidity on Photochemical Oxidant Production," A.C.S. Annual Meeting, Division of Water and Waste Chemistry, Los Angeles, Calif., 1963.
4. B. Dimitriades, "Methodology in air pollution studies using irradiation chambers," *J. Air Poll. Control Assoc.* **17**, 460 (1967).
5. J. J. Bufalini and A. P., Altshuller, "Oxidation of nitric oxide in the presence of ultraviolet light and hydrocarbons," *Environ. Sci. Technol.* **3**, 469 (1969).
6. W. E. Wilson, Jr., and A. Levy, "A Study of Sulfur Dioxide in Photochemical Smog," Battelle Memorial Institute Report to American Petroleum Institute Project S11, Columbus, Ohio, 1968.
7. W. E. Wilson, Jr. Personal Communications. Battelle Memorial Institute, Columbus, Ohio, 1969
8. A. E. O'Keefe and G. Ortman, "Primary standards for trace gas analysis," *Anal. Chem.* **38**, 760 (1966).

9. H. Scheffe, *The Analysis of Variance*, New York, John Wiley & Sons, Inc., 1963, p. 73.

10. D. W. G. Style, *Photochemistry*, London, England, Methuen and Co., 1930, p. 36.

11. L. F. Wayne, R. J. Bryan, M. Weisburd, and R. Danchick, "Comprehensive Technical Report on All Atmospheric Contaminants Associated with Photochemical Air Pollution," TM-(L)-4411/002/01 Report on Contract No. CPA 22-69-108 to N.A.P.C.A., C.P.E., P.H.S., Dept. of Health, Education, and Welfare, System Development Corp., Santa Monica, Calif., 1970.

12. A. Goetz and R. Pueschell, "Basic mechanisms of photochemical aerosol formation," *Atmos. Environ.* **1**, 287 (1967).

13. C. Walling, *Free Radicals in Solution*, New York, John Wiley & Sons, Inc., 1957. pp. 35.

14. A. Goetz, and O. J. Klejnot, "The Aerocolloid Formation of Reactive Hydrocarbons in the Biosphere," The International Conference on Cloud Physics," Toronto, Canada, 1968.

15. C. E. Decker, and W. W. Page, "A Photochemical Study of Systems Containing Blends of Hexene-1, Nitrogen Dioxide, and Sulfur Dioxide," Unpublished M.S. Thesis, Dept. of Environmental Sciences and Engineering, Univ. of North Carolina, Chapel Hill, N. C., 1965.

16. F. W. Went, "Organic matter in the atmosphere and its possible relation to petroleum formation," *Proc. Natl. Acad. Sci. U.S.A.*, **46**, 212 (1960).

17. F. W. Went, "On the nature of Aitken condensation nuclei," *Tellus* **28**, 549 (1966).

18. L. A. Ripperton, "Ozone and ozone precursors in the atmosphere of Chapel Hill, N. C.," *J. Geophys. Res.* **70**, 5009 (1965).

19. P. K. Mueller, N. O. Transah, Y. Tokiwa, and E. L. Kothny, "Series vs. Parallel Continuous Analysis for NO, NO_2, and NO_x," Ninth Conference on Methods in Air Pollution and Industrial Hygiene Studies, Pasadena, Calif., 1968.

20. H. E. Jeffries and O. White, "Alphapinene and Ozone: Some Atmospheric and Biospheric Implications," Masters Thesis, Dept. Environ. Sci. and Eng., Univ. North Carolina, 1967.

Oxidation of Atmospheric SO_2 by Products of the Ozone–Olefin Reaction

R. A. COX
S. A. PENKETT

OZONE is the most abundant oxidant in the troposphere and is generated in the stratosphere by the reaction of oxygen atoms, formed by photolysis, with molecular oxygen. Some is then transferred to the troposphere and 0–0.05 p.p.m. of ozone is found in unpolluted air at ground level[1]. Ozone can also be formed directly in tropospheric polluted air by the photolysis of NO_2 in the presence of hydrocarbons; concentrations in excess of 0.15 p.p.m. are regularly recorded in Los Angeles[2] and other cities of the United States. Significant increases in equivalent ozone concentrations over background levels have been recorded in Holland[3]—up to 0.10 p.p.m. in conditions suitable for the build up of photochemical oxidants. The ambient sulphur dioxide level was found to vary inversely with equivalent ozone level.

The reaction between ozone and SO_2 is too slow to be significant at atmospheric concentrations[4], but the reaction with unsaturated hydrocarbons is very rapid even at concentrations less than 1 p.p.m. (ref. 5). The products of this reaction include a highly reactive species which has a strong germicidal[6] and phytotoxic[7] effect, and it has been suggested that the active species is a peroxide zwitterion formed during the ozone–olefin reaction[8,9]

$$R_2C=CR_2+O_3\rightarrow R_2C^+OO^-+R_2CO \qquad (1)$$

Such a species, with a pronounced charge separation, should also be an effective oxidant for the conversion of SO_2 to SO_3.

We report here experiments which demonstrate such an oxidation; they were carried out in a 220 l. aluminium chamber[10,11] which was flushed out with a stream of purified air before the reactants were introduced. A slow air stream (about 1 l./min^{-1}) also vented the chamber during runs. Relative humidity was 40% in all experiments. Ozone was generated externally by passing 200 cm^3 min^{-1} of cylinder oxygen through a small aluminium box containing mercury vapour ultraviolet lamps. Ozone levels of about 1 p.p.m. were obtained in the chamber after generation for 2 min. The SO_2 was measured into a small glass cell on a vacuum system and injected into the chamber. The reaction was started by injecting the required amount of liquid olefin into the venting air stream, where it vaporized before entering the chamber.

A small fan was installed close to the injection point to ensure rapid mixing.

The ozone concentration was monitored continuously using a Brewer cell[12] containing buffered KI solution. The output current was monitored by measuring the potential difference across a 500Ω series resistor. The response time of this system was, however, too slow to record accurately the changing ozone levels in the more rapid reactions. Also SO_2 and oxidant formed during the reaction interfered with the measured values. Hydrocarbon concentrations were measured with a gas chromatograph fitted with a flame ionization detector.

The measurement of SO_2 and its oxidation products using radioactive $^{35}SO_2$ has been discussed previously[10]. The product of the oxidation, an aerosol, is separated from the gaseous SO_2 on a membrane filter. Both the filters and samples of gaseous SO_2 collected in H_2O_2 solution are assayed by liquid scintillation counting and the mass of sulphur calculated from the count rate.

We first studied the ozone–olefin reaction in the absence of SO_2; the reaction followed second order kinetics and approximately equal amounts of ozone and olefin were consumed. The second order rate constants, k_1, for each olefin agreed quite well with previous values[5].

Large amounts of aerosol were produced when SO_2 was added to the mixture, and because the formation pattern was similar to that produced photochemically[10,11] we assumed that the aerosol was sulphuric acid. The SO_2 had no significant effect on the rate of disappearance of ozone or hydrocarbon but a small increase in the carbonyl product yields was observed.

Fig. 1 shows the build up of sulphuric acid aerosol (expressed as p.p.m. of SO_3) for four different olefins. In each case the initial reactant concentrations were about 2 p.p.m. olefin, 0.4 p.p.m. ozone, and 0.11 p.p.m. SO_2. There was no observable induction period after the addition of olefin to the SO_2–ozone mixture and aerosol formation ceased when the ozone concentration had decreased to zero. The amounts of aerosol formed were similar for all the olefins used and were considerably less than the amount of ozone consumed. Table 1 shows that, for the ozone-4-methyl-1-pentene reaction, the amount of aerosol formed only increased by a factor of 3 when the SO_2 concentration was increased twenty-fold. The initial rate of aerosol production also does not vary with SO_2 concentration. The reaction kinetics are complex and the difference between the amount of ozone consumed and the amount of aerosol formed is not constant during the reaction.

Fig. 1 also shows that there is a variation in the initial rates of aerosol production for the different olefins. This variation is expressed numerically in Table 2 which contains data collected

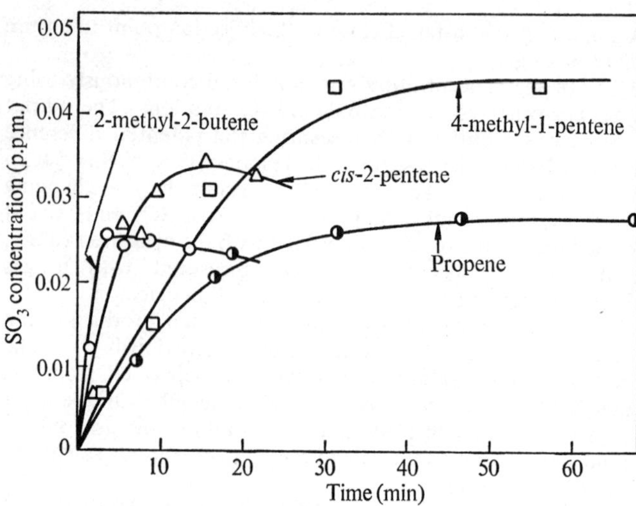

Fig. 1 Sulphuric acid aerosol formation during the oxidation of SO_2 (0.11 p.p.m.) in the presence of O_3 (0.4 p.p.m.) and olefins (2 p.p.m.).

Table 1 Variation of the Amount of Aerosol formed by the Ozone-4-methyl-1-pentene Reaction with Initial SO_2 Concentration

Initial SO_2 concentration (p.p.m.)	Maximum aerosol concentration (p.p.m. SO_3)	Ratio $\dfrac{SO_3 \text{ (max)}}{O_3 \text{ (consumed)}}$	Initial d/dt (SO_3) (p.p.m./min^{-1})
0.058	0.021	0.0525	0.0030
0.108	0.045	0.112	0.0029
1.19	0.060	0.150	0.0030

Initial olefin and ozone concentration was 2.0 4.0 p.p.m. respectively.

Table 2 Rate of Aerosol Formation for Various Olefins

Olefin	Initial rates (p.p.m. min^{-1}) $-d/dt\,(O_3)$	Initial rates (p.p.m. min^{-1}) $d/dt\,(SO_3)$	$\dfrac{SO_3 \text{ (max)}}{O_3 \text{ (consumed)}}$
2-Methyl-2-butene	0.34	>0.02	0.063
cis-2-Pentene	0.20	0.017	0.085
4-Methyl-1-pentene	0.013	0.0029	0.112
Propene	0.016	0.0021	0.070

Initial concentration was 2.0 p.p.m. olefin, 0.4 p.p.m. ozone and 0.11 p.p.m. SO_2.

in the presence and absence of SO_2. Changes in the initial rate of aerosol formation follow the changes in the rate of disappearance of ozone (determined in the absence of SO_2) quite closely. The rates of aerosol formation are much greater for the internally unsaturated olefins, *cis*-2-pentene and 2-methyl-2-butene, than for the terminally unsaturated propene and 4-methyl-1-pentene.

It is interesting to speculate whether these same oxidants are responsible for the observed photo-oxidation of SO_2 in the presence of oxides of nitrogen and olefins[11] because the formation of ozone and aerosol in the latter system shows similar features. Formation of sulphate aerosol does not occur during irradiation of a mixture of NO, olefin and SO_2 until the NO has been converted to NO_2 (ref. 13) and this also applies to the photochemical formation of ozone from nitric oxide and hydrocarbons. Similarly, the maximum aerosol yield coincides with the maximum yields of ozone and peroxyacyl nitrates when the ratio of initial concentrations of olefin to NO_2 is varied in the photolysis of olefin, SO_2 and NO_2 mixtures[14].

The concentration of ozone necessary to account for the photolytic rates of aerosol formation observed in our previous experiments[11] can be calculated from d/dt (SO_3) (Table 2). With initial concentrations of 0.33 p.p.m. NO, 0.11 p.p.m. SO_2 and 0.56 p.p.m. *cis*-2-pentene, the maximum rate of aerosol production $(5.95 \times 10^{-4}$ p.p.m. $min^{-1})$ would require an ozone concentration of 0.112 p.p.m., assuming that the hydrocarbon concentration had dropped to 0.25 p.p.m. when the maximum rate was reached. In the case of propene an ozone concentration of 0.026 p.p.m. would be required to account for the maximum rate of aerosol production $(1.37 \times 10^{-5}$ p.p.m. $min^{-1})$ with a mixture initially containing 0.4 p.p.m. NO and propene and 0.125 p.p.m. SO_2. Photo-oxidation data for NO–olefin mixtures[15] show that ozone concentrations of this magnitude are certainly produced in this system, and may well account for the formation of sulphuric acid aerosols during photo-oxidation in the presence of SO_2.

The general occurrence in Britain of a germicidal agent in the air has been demonstrated recently[16,17] and the properties of this so-called open air factor (OAF) are similar to the biologically active products of ozone–olefin reactions. OAF seems to be more intense in rural air than in city air and, although independent of the ozone concentration, is thought to be associated with a reaction between ozone and the olefinic constituents of vehicle exhaust gases[17]. The effects of the ozone–olefin reaction seem to be widespread and could make a significant contribution to the oxidation of sulphur dioxide in the atmosphere. At ozone and olefin concentrations of

0.05 p.p.m., the oxidation rate of 0.1 p.p.m. SO_2 is calculated to be about $3\% h^{-1}$ for *cis*-2-pentene and $0.4\% h^{-1}$ for propene.

[1] Junge, C. E., *Air Chemistry and Radioactivity*, 37 (Academic Press, New York and London, 1963).
[2] *Air Quality Data from the National Air Sampling Networks, 1964–1965* (US Department of Health, Education and Welfare, 1966).
[3] Wisse, J. A., and Velds, C. A., *Atmos. Environ.*, **4**, 79 (1970).
[4] Cadle, R. D., in *Air Pollution Handbook* (edit. by Magill, P. L., Holden, F. R., and Ackley, C.), 3 (McGraw-Hill, New York, 1956).
[5] Leighton, P. A., *Photochemistry of Air Pollution*, 159 (Academic Press, New York, 1961).
[6] Druett, H. A., and Packmann, L. P., *Nature*, **218**, 699 (1968),
[7] Darley, E. F., Stephens, E. R., Middleton, J. T., and Hanst. P. L., *Intern. J. Air Pollution*, **1**, 155 (1959).
[8] Dark, F. A., and Nash, T., *J. Hyg., Camb.*, **68**, 245 (1970).
[9] Arnold, W. N., *Intern. J. Air Pollution*, **2**, 167 (1959).
[10] Cox, R. A., and Penkett, S. A., *Atmos. Environ.*, **4**, 425 (1970).
[11] Cox, R. A., and Penkett, S. A., *Nature*, **229**, 486 (1971).
[12] Brewer, A. W., and Milford, J. R., *Proc. Roy. Soc.*, A, **256**, 470 (1960).
[13] Wilson, jun., W. E., and Levy, A., *J. Air Poll. Control Assoc.*, **20**, 385 (1970).
[14] Altshuller, A. P., and Bufalini, J. J., *Photochem. Photobiol.*, **4**, 97 (1965).
[15] Altshuller, A. P., Kopczynski, S. L., Lonneman, W. A., Becker, T. L., and Slater, R., *Environ. Sci. Tech.*, **1**, 899 (1967).
[16] Druett, H. A., and May, K. R., *Nature*, **220**, 395 (1968).
[17] Druett, H. A., and May, K. R., *New Scientist*, March 13 (1969).

Electron Spin Resonance and Optical Studies of the Interaction between NO_2 and Unsaturated Lipid Components[1]

R. M. ESTEFAN, E. M. GAUSE, AND J. R. ROWLANDS

Electron spin resonance and optical spectroscopic studies of the reactions between nitrogen dioxide and selected unsaturated phospholipids, fatty acids, and hydrocarbons are reported. The chemical natures of both the transient free radicals and stable free radicals produced in these reactions are discussed.

INTRODUCTION

As a result of increased interest and research into the biological effects of both air pollutants and tobacco smoke, there has emerged an increased awareness of the potentially hazardous properties of the oxides of nitrogen.

Arioka (1967) has recently demonstrated that pulmonary damage was experimentally induced in rats by inhalation of NO_2. He successfully demonstrated a relationship between the mortality rate and the NO_2 concentration, and also showed that the blood of the exposed rats contained an elevated concentration of both metmyoglobin (4–6%) and potassium ions. Symon (1965) has reported that elevated levels of metmyoglobin (3–5%) have been observed in children living in an environment containing high concentrations of oxides of nitrogen. Rowlands *et al.* (1968) have demonstrated by electron spin resonance studies that constituents of tobacco smoke pass through the lung membrane and enter the bloodstream to give rise to four types of paramagnetic iron complexes, of which one complex was positively identified as being produced by oxides of nitrogen contained in the tobacco smoke. These results were obtained both on blood from rats and mice subjected to the smoke from two cigarettes and from experiments in which the blood of a rabbit was pumped through the lung of a rabbit while the lung was mechanically forced to breathe and "smoke" cigarettes. It was also reported in a preliminary manner by these latter authors that nitric oxide produces a free radical species upon interaction with the lung phospholipid phosphatidyl ethanolamine.

Recent studies by Thomas *et al.* (1967) concerned with the effects of NO_2 on young rats, have demonstrated that the inhalation of 1 ppm for one hour or 0.5 ppm for four hours leads to significant morphological changes in their lung-mast cells. The morphological changes are thought to occur both as a result of the chemical interaction of NO_2 with the cell membrane and as a result

[1] These studies were supported by a grant from The Council for Tobacco Research (U. S. A.).

of the interaction of NO_2 with certain specific but undefined cellular elements which lead to a loss in cell integrity.

Felmeister *et al.* (1968) have reported on surface pressure measurements on films of both saturated and unsaturated phospholipids. They find that, whereas no observable interactions were observed in the case of the saturated phospholipids, films of egg lecithin, an unsaturated phospholipid, showed significant changes in the surface pressure-surface area curves in the presence of all atmospheres studied containing nitrogen dioxide. They attribute their observed effects to a chemical interaction of NO_2 with the double bonds of the lecithin.

Thomas *et al.* (1968) have recently reported absorption spectroscopic evidence of lipid peroxidation from extracts of the lungs of rats which have been exposed to one part per million atmospheres of NO_2 for both single periods of four hours and repeated four-hour exposures over a period of six days. Their results suggested that the extent of peroxidation was cumulative. These authors also report that structural changes are induced in the lung lipids upon exposure to the nitrogen dioxide.

The evidence accumulated to date suggests that NO_2, particularly, exerts its biological effects by: (1) reacting with unsaturated bonds of physiological lipid materials (Felmeister *et al.*, 1968; Thomas *et al.*, 1967), and (2) complexing heme iron or otherwise interacting with iron-containing components of blood (Rowlands *et al.*, 1968; Symon, 1965; Arioka, 1967). The widespread occurrence of lipids in biological tissues and the signicance of the biological roles of various lipids makes it extremely important to understand the mechanism of the interaction of the oxides of nitrogen with these substances.

In this communication, we wish to present results obtained in an electron spin resonance (ESR) and optical study of free radicals produced by the interaction of NO_2 with an unsaturated fatty acid (oleic acid), its ester (methyl oleate), two phospholipids (phosphatidyl ethanolamine and L-α-lecithin), and two unsaturated aliphatic hydrocarbons chosen on the basis of location and configuration of the double bond site.

EXPERIMENTAL PROCEDURES

Materials and Methods

Phosphatidyl ethanolamine and lecithin (90%, bovine) were obtained from Nutritional Biochemical Corporation; oleic acid, 1-octadecene and methyl oleate from Aldrich Chemical; and trans-5-decene from K & K Laboratories. Chloroform, dimethoxyethane, and dioxane were used as solvents during the course of the studies. All three were found to be both suitable as solvents for the phospholipids, fatty acids, and hydrocarbons used in this study, and to be relatively inert to nitrogen dioxide.

Electron spin resonance measurements of the stable free radicals were made using a Varian V-4500 X-band spectrometer utilizing 100 Kc/sec modulation. Measurements of g factors were made by reference to the known g factor of $g = 2.0036$ of diphenyl picryl hydrazyl, and hyperfine coupling constants were measured by comparison with measurements of the nitrogen hyperfine coupling constants of Fremy's salt. Measurements of the transient free radicals were made utilizing a Jeolco P10 radical detector with flow cell accessory. Measurements of the optical spectra of reaction mixtures were made with a Cary 14 UV visible and infrared spectrometer.

All measurements were made at room temperature ($\sim$20–25°C), unless specified otherwise.

Results

Solutions of the phospholipids, phosphatidyl ethanolamine and lecithin; the fatty acid, oleic acid; and the methyl ester, methyl oleate; together with the unsaturated hydrocarbons octadecene and trans-5-decene, were all shown by electron spin resonance to give rise to stable free radicals upon reaction with solutions of NO_2. In addition to the stable free radicals, it was established that the reaction of dimethoxyethane solutions of oleic acid and a DME solution of NO_2 gave rise to a transient free radical which decayed to yield one of the stable free radicals. The stable free radicals were radicals in which the unpaired electron was predominantly located on one nitrogen atom, resulting in a characteristic three-line ESR spectrum of 1:1:1 intensities. In the case of phosphatidyl ethanolamine, in addition to the nitrogen radicals, a free radical with no observable hyperfine structure was formed. In the following discussion, the nitrogen radicals will be classified into three types by the magnitude of the observed nitrogen hyperfine coupling constants. Type I radicals give rise to nitrogen hyperfine coupling constants of *ca* 7 gauss, Type II *ca* 12–15 gauss, and Type III *ca* 29 gauss. A summary of the hyperfine coupling constants and *g* factors for the stable free radicals observed for each of the compounds studied are listed in Table 1.

The relative concentration of each type of radical was found to be dependent both upon the concentration of lipids or hydrocarbons and nitrogen dioxide and to be influenced by photochemical processes. The details of the concentration dependencies appear to be quite complex. Preliminary experiments conducted for both oleic acid and octadecene suggest that, at the most dilute concentrations studied, Type I radicals predominate, whereas both at the higher concentrations and in the presence of UV radiation, Types II and III become equally important. However, the principal purpose of this communication is to report on the types of free radicals that are produced in reactions of NO_2 with unsaturated hydrocarbon chains, rather than detailed kinetics of the various reactions. Details of the spectra obtained for the individual compounds studied are as follows.

Phosphatidyl ethanolamine and lecithin. Solutions (0.1 M) of phosphatidyl ethanolamine and L-α-lecithin were dissolved in chloroform and 1 ml of the resulting solutions were mixed with 0.15 ml of chloroform saturated with NO_2. The electron spin resonance spectra of the resulting solutions are illustrated in Figs. 1 and 2. The spectra consist of a superposition of three types of radicals, each of which exhibits hyperfine interaction to a single nitrogen nucleus. In addition to the nitrogen radicals, the phosphatidyl ethanolamine reaction gives rise to a free radical in which no evidence of hyperfine structure is observed. The

TABLE 1
ESR CHARACTERISTICS OF RADICALS PRODUCED FROM REACTION OF NO$_2$ WITH VARIOUS UNSATURATED LIPIDS

Compound[a]	Solvent	Type I a_N (gauss)	Type I g-factor	Type II a_N	Type II g-factor	Type III a_N	Type III g-factor
L-α-lecithin (bovine)	CHCl$_3$	7.12	2.0061	11.48	2.0065	29.30	2.0053
				11.01	2.0061		
Phosphatidyl[b] Ethanolamine	CHCl$_3$	7.12	2.0069	11.66	2.0066	29.03	2.0054
				10.7	2.0064		
Oleic acid[c]	CHCl$_3$	6.88	2.0068	11.62	2.0064	29.07	2.0050
	dioxane	7.29		14.74			
	dioxane	7.10		14.07			
	dioxane/DME	6.97					
Methyl oleate	CHCl$_3$	6.88	2.0064	12.24	2.0063	29.07	2.0054
	dioxane	7.24		15.01			
1-Octadecene	DME	6.76	2.0073	12.48	2.0062	29.37	2.0056
	CHCl$_3$			13.15			
	dioxane	7.24		14.20			
Trans-5-decene		7.29	2.0070	12.48	2.0062	29.12	2.0060

[a] All obtained at room temperature.

[b] An additional single line was observed at $g = 2.0069$.

[c] An initial radical was observed under flow conditions. See text.

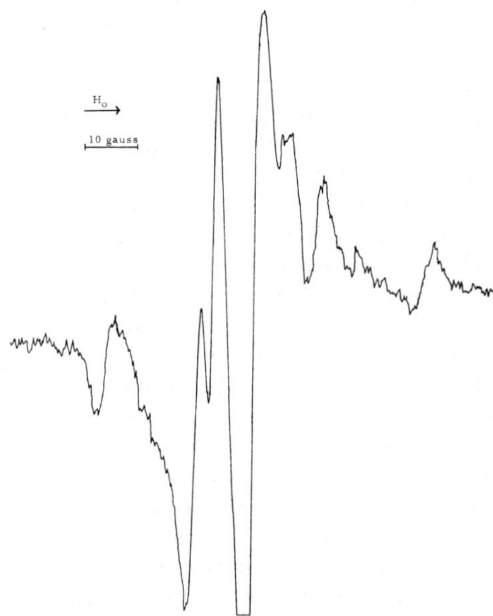

FIG. 1. ESR spectrum of 0.1 M phosphatidyl ethanolamine in chloroform mixed with 0.15 ml of NO -saturated chloroform.

g factor for the single line species is $g = 2.0069$; coupling constants and g factors of the nitrogen radicals are contained in Table 1.

Oleic acid and methyl oleate. Both oleic acid and methyl oleate solutions in chloroform, dimethoxyethane and dioxane were found to give rise to the three types of nitrogen radicals upon reaction with nitrogen dioxide. A typical electron spin resonance spectrum in which all three types are present is illustrated in Fig. 3. The free radical giving rise to the nitrogen hyperfine coupling of 29 gauss was found to be produced only at the higher NO_2 concentrations utilized. The absolute magnitude of the nitrogen coupling constants of both Type I and Type II radicals were found to be solvent dependent. Coupling constants and g factors for these species measured in several solvents are tabulated in Table 1.

The reaction between oleic acid and nitrogen dioxide was also studied in a flow system in an effort to establish the nature of the initial reaction between the reactants. In these experiments, solutions of oleic acid dissolved in dioxane (20% by volume) and dimethoxyethane solutions of NO_2 (estimated to be approximately 25% saturated) were contained in separate arms of a standard Jeolco flow cell apparatus. No effort was made to exclude air in the preparation of reactant solutions. Under flow conditions of approximately 0.5 ml/sec of each reactant, the reaction mixture was seen to give rise to a complex free radical signal with hyperfine structure extending over 36 gauss. A typical spectrum of this species is shown in Fig. 4a. Upon stopping the flow of the reactants, the spectrum of this initial radical was found to be rapidly replaced by the triplet characteristic of Type I radicals (Fig. 4b). Although we have not directly measured the decay time of the initial species, our indirect evidence is that it occurs rapidly, since flow experiments in which the reactants were allowed to flow at a rate of approximately 0.25 ml/sec were found to give rise to the Type I free radical spectrum. Under higher resolution, each component of the triplet of the Type I radical can be observed to be further split into a triplet of 1:2:1 relative intensities (Fig. 5), and of 1.58-gauss coupling.

At the concentrations used for the above experiments, there was no evidence for the formation of the Type II radical species. Since the concentrations used were not significantly different from those used for the preparation of the spectrum illustrated in Fig. 3, the possibility that the formation of Type II radicals in this system is photochemically induced was considered.

In order to investigate this possibility, flow experiments were conducted using the same concentration solutions that gave rise to the spectra illustrated in Figs. 4a and 4b, while the sample cell in the cavity was kept under continuous illumination with the light from a medium pressure, Hanovia 140-watt UV source. No significant differences were observed in the electron spin resonance spectra obtained under flow, but upon stopping the flow, the electron spin resonance spectrum illustrated in Fig. 6 was obtained. This spectrum can be seen to consist of a superposition of the characteristic triplet spectra of the Type I and Type II radicals. It thus appears from these experiments that the formation of Type II radicals is enhanced by the action of ultraviolet radiation.

1-Octadecene. When nitrogen dioxide is bubbled into 1-octadecene dissolved in DME, or when DME solutions of both reactants are mixed, all three stable

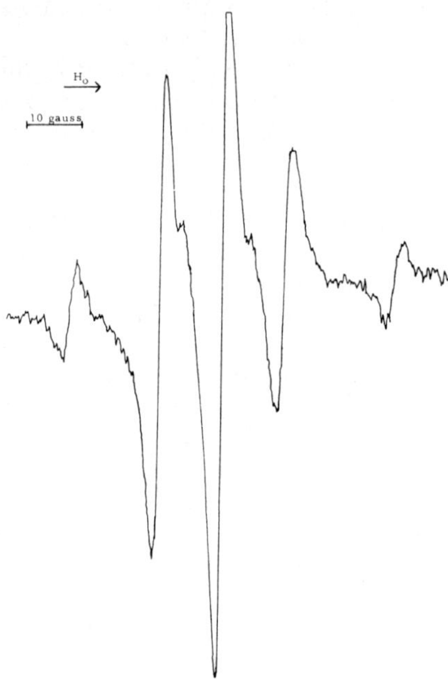

FIG. 2. ESR spectrum of 0.1 M L-α-lecithin in chloroform mixed with 0.15 ml of NO₂-saturated chloroform.

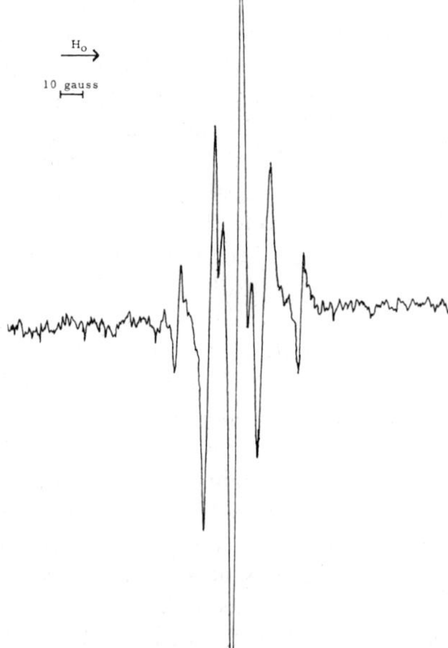

FIG. 3. ESR spectrum of 50% (vol) methyl oleate in CHCl₃ upon exposure to gaseous NO₂.

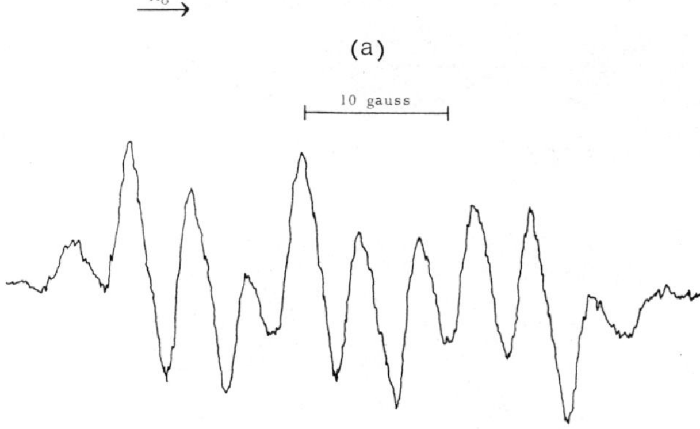

Fig. 4a.   ESR spectrum of radical produced under continuous flow conditions in reaction of 20% (vol) oleic acid in dioxane with NO_2 in dimethoxyethane.

radical species are obtained (Fig. 7). Coupling constants and g factors for each of the species are included in Table 1. In agreement with the observations made on oleic acid, the Type III radical is again found to be less concentrated than the other two species and to be present only under conditions in which Type II radicals are present in high concentration.

In flow experiments, we have so far been unable to observe an initial free radical species comparable to that observed in the oleic acid experiments. Our

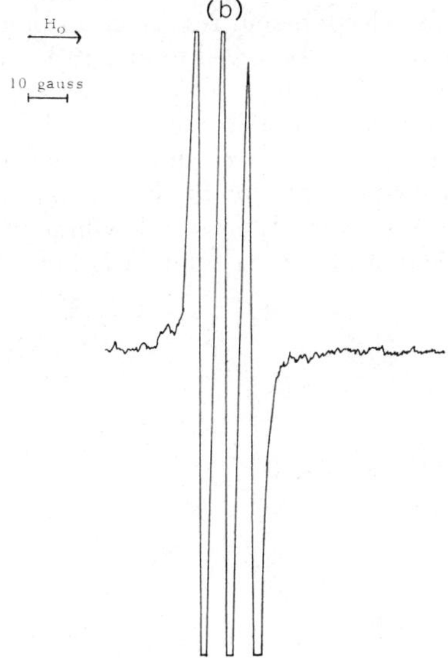

Fig. 4b.   ESR spectrum of Type I radical formed upon stopping flow in oleic acid-NO_2 reaction.

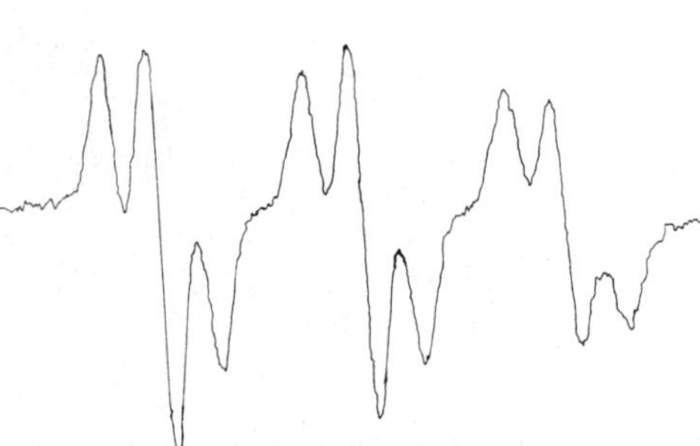

Fig. 5. ESR spectrum of Type I radical of Fig. 4b under higher resolution.

inability to observe such a species appears to be due to decreased lifetime of such a species rather than its absence in the reaction, since a low concentration of Type I radicals was observed under the fastest flow conditions that we could attain. Upon stopping the flow of the reactants, the concentration of the Type I species was found to increase rapidly. Just as observed in the flow experiments with oleic acid, the production of Type II radicals in the octadecene system is apparently enhanced by UV light.

A characteristic of the octadecene/NO_2 solutions in which the Type II free radicals were shown by electron spin resonance to predominate was their blue color. These solutions were also observed to give rise to a white precipitate upon standing. Optical spectra of the blue solutions were measured and were shown to give rise to two distinct absorption regions. A characteristic absorption was observed in these solutions with the maximum absorption occurring at 6600 Å. A further absorption characteristic of these solutions was an absorption band extending from 3800 Å to 3200 Å, in which vibrational structure was plainly visible. A typical spectrum of such a solution is illustrated in Fig. 8. Attempts

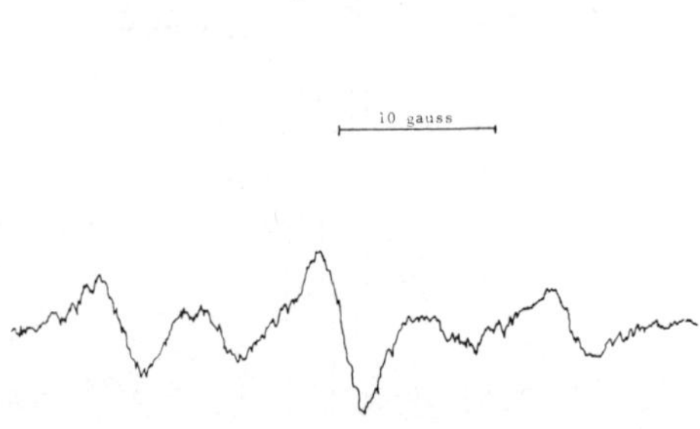

Fig. 6. ESR spectrum of stopped-flow radicals produced in oleic acid-NO_2 system under UV illumination.

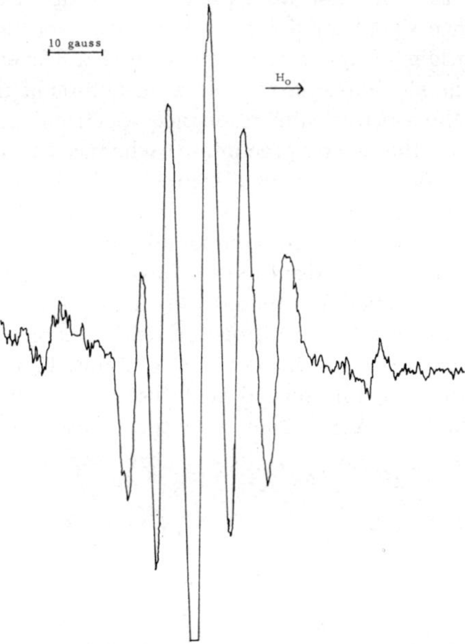

FIG. 7. ESR spectrum of octadecene in dimethoxyethane upon exposure to gaseous NO₂.

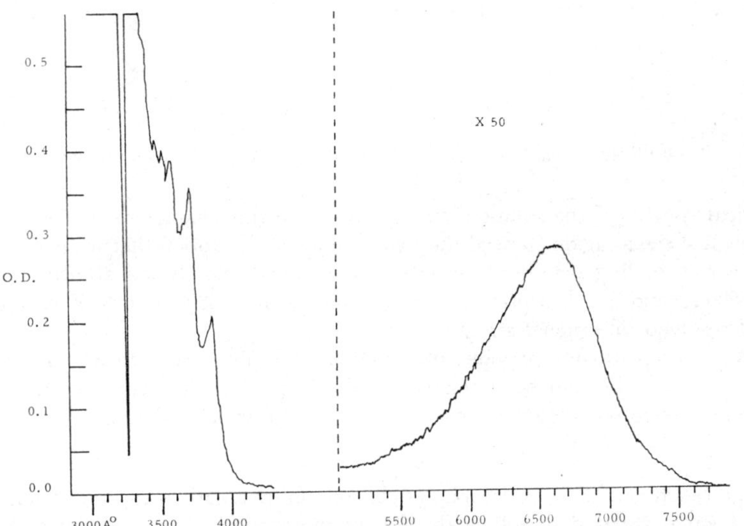

FIG. 8. Optical spectra of octadecene in dimethoxyethane upon exposure to gaseous NO₂.

to correlate the intensity of these absorption bands with the amplitude of the electron spin resonance signal for the Type II radical were unsuccessful. It was observed that the origin of the failure to observe a correlation between the optical spectra and the electron spin resonance amplitude of the Type II radicals lay in the fact that the electron spin resonance spectrum increased in intensity with the formation of the white precipitate, whereas the optical spectra decreased in intensity upon formation of the precipitate. In separate experiments, the white precipitate was collected, washed thoroughly in *n*-hexane, and air-dried. Weighed samples of the white compound were examined for free radical content and were shown to give rise to approximately 1.0×10^{17} spins per gram.

Trans-5-decene. Both dimethoxyethane and chloroform solutions of trans-5-decene, when allowed to react with nitrogen dioxide gas or mixed with solutions of nitrogen dioxide, gave rise to blue-colored solutions. Electron spin resonance studies of such solutions again showed that the same three types of nitrogen radicals had been formed. A representative spectrum is illustrated in Fig. 9.

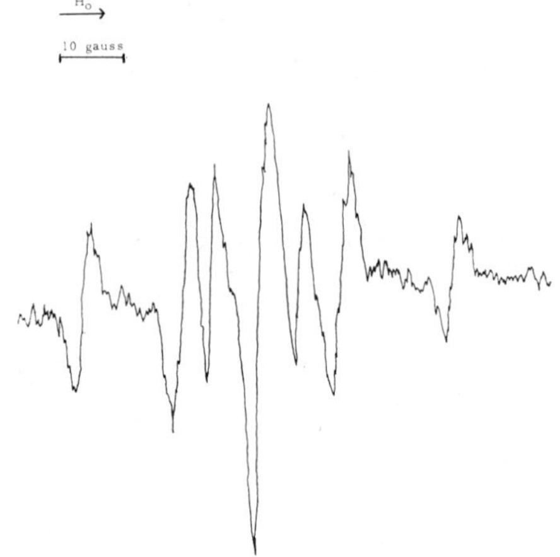

FIG. 9. ESR spectrum of trans-5-decene in dimethoxyethane upon exposure to gaseous NO_2.

Optical spectra of the solutions that gave rise to this characteristic electron spin resonance signal again showed the two regions of absorption that were also noted for corresponding octadecene solutions, i.e., the rather broad absorption band at 6600 Å, and an absorption system extending from 3800 Å to 3350 Å, showing well resolved vibrational structure.

In accord with our previous observations, the Type II radical concentration was shown to be enhanced relative to the Type I radical concentration when reactions were conducted in the presence of ultraviolet radiation.

DISCUSSION

The reaction of NO_2 with compounds containing a carbon-carbon double bond has been postulated to follow the following reaction scheme (Levy and Rose, 1947):

$$RCH{=}CHR + NO_2 \rightarrow RCHCHNO_2R. \tag{1}$$

$$RCHCHNO_2R + NO_2 \rightarrow RCHNO_2{-}CHNO_2R. \tag{2}$$

The electron spin resonance spectrum of the transient free radical observed in the flow cell measurements of the oleic acid-NO_2 reaction in dimethoxyethane may be readily interpreted in terms of reaction 1. The ESR spectrum of this radical, illustrated in Fig. 4a, is consistent with a free radical in which the unpaired electron exhibits hyperfine interaction with four hydrogen atoms with coupling constants of $A_{H_1} = 16$ gauss, $A_{H_2} = 4$ gauss, $A_{H_3} = 12$ gauss, $A_{H_4} = 4$ gauss. Immediately after stopping the flow of the reactants through the cavity, the electron spin resonance spectrum of this initial radical was shown to have disappeared and to have been replaced by a Type I free radical (Fig. 4b). The principal features of the electron spin resonance spectrum of this species consist of a triplet of equal intensity lines consistent with the interaction of the unpaired electron with a nitrogen nucleus. Under higher resolution (Fig. 5), each component of the triplet was seen to be further into a triplet of 1:2:1 relative intensity, and 1.58 gauss coupling. The formation of this species from the initial radical most probably occurs as a result of the following intramolecular hydrogen atom transfer reaction:

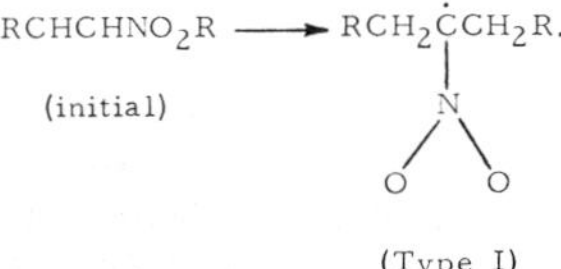

(Type I)

Spin density distributions for the $> C\text{-}NO_2$ fragment calculated by simple Hückel molecular orbital theory are included in Table 2, together with nitrogen hyperfine coupling constants calculated using the Rieger and Fraenkel (1963) relationship:

$$a_N = (S^N + Q_{NO}^N + 2Q_{NO}^N)\rho^N + Q_{CN}^N\rho^C + 2Q_{ON}^N\rho^O.$$

In this expression, ρ^N, ρ^O, ρ^C are respectively the π electron spin densities on the nitrogen, oxygen, and carbon atoms; S^N represents the contribution to the splitting from the nitrogen $1s$ electrons and the Q terms account for the contributions of the $2s$ electrons where, for example, Q_{ON}^N is the sigma-pi parameter for the nitrogen nucleus resulting from the interaction between the ON bond and the π electron spin density on the oxygen atom. Values of the parameters used, taken from estimates made by Gross and Symons (1966), were

$$(S^N + Q_{NC}^C + 2Q_{NO}^N) = 27 \text{ gauss,}$$
$$Q_{NO}^N = -3.7 \text{ gauss,}$$
$$Q_{CN}^N = -5 \text{ gauss.}$$

In view of the approximate nature of the Hückel molecular orbital method, the agreement between the calculated and observed values of the nitrogen hyperfine coupling constant is good. However, of more importance is the fact that the

TABLE 2

CALCULATED SPIN DENSITY DISTRIBUTIONS AND NITROGEN
COUPLING CONSTANTS FOR $>$C—NO_2 FRAGMENT

	(1)	(2)		$a_N(1)$	$a_N(2)$	$a_H(1)$	$a_H(2)$
			Gauss				
ρ_C	.3306	.2310					
ρ_N	.3352	.3170		6.17	5.73	$\sim$12	$\sim$9
ρ_O	.1664	.2256					

(1) $\alpha_N = \alpha_C + 1\beta$, $\beta_{CN} = 1\beta$;

$\quad \alpha_O = \alpha_C + 2\beta$, $\beta_{N-O} = 0.7\beta$.

(2) $\alpha_N = \alpha_C + 1.5\beta$, $\beta_{CN} = 1\beta$;

$\quad \alpha_O = \alpha_C + 2\beta$, $\beta_{NO} = 0.7\beta$.

theory does predict that the delocalization of the unpaired electron is sufficiently extensive to allow a relatively low residual spin density at the carbon atom. In the last column of Table 2, hyperfine coupling constants for the methylene protons are calculated for a configuration in which one proton from each methylene group projects onto the nodal plane of the carbon $2p$ orbital containing the unpaired electron. For this configuration, projections of the second proton from each methylene group onto this plane make an angle of 30° to the p orbital direction. The β hydrogen coupling constants are then calculated from the well-known $B \cos^2 \theta$ relationship using a value of the constant B of 50 gauss for a spin density of 1.0. The agreement with experiment, although less than satisfactory, does predict a considerable reduction of the magnitude of the β hydrogen coupling constants from the value of 40 gauss, which would have been the predicted coupling constant for this configuration for a spin density of 1.0 on the carbon. The agreement with experiment would be expected to be improved by a more sophisticated level of approximation than that afforded by the Hückel molecular orbital method.

Type II free radicals, which were shown to be produced in all systems studied, were produced in greatest concentration relative to the Type I radicals both at higher nitrogen dioxide concentrations studied and in the presence of ultraviolet radiation. Free radicals with nitrogen hyperfine coupling constants comparable to Type II radicals have been previously reported to be produced in the reaction of nitrogen dioxide with methyl methacrylate (McRae and Symons, 1966). The formation of these radicals was considered to proceed according to the following general scheme:

$$2R_1R_2C = CR_3R_4 + NO_2$$

$$\rightarrow \quad
\begin{array}{c}
R_1R_2{-}C{-}C{-}R_3R_4 \\
\diagdown \ \diagup \\
O{-}N \qquad O \\
\diagdown \ \diagup \\
R_1R_2C{-}CR_3R_4
\end{array}
\quad + \quad
\begin{array}{c}
R_1R_2C{-}CR_3R_4 \\
\diagdown \ \diagup \\
O{-}N \qquad O \\
\diagdown \ \diagup \\
R_3R_4C{-}CR_1R_2
\end{array}$$

to yield nitroxide free radicals.

Such radicals have been shown to display characteristic nitrogen hyperfine couplings in their electron spin resonance spectra of $10 \rightarrow 16$ gauss dependent upon the chemical nature of the substituent groups R_i. They are planar radicals with the unpaired electron in a π molecular orbital which is predominantly localized on the nitrogen atom. The formation of Type II radicals according to the above reaction scheme for the reaction between NO_2 and oleic acid or trans-5-decene would give rise to the following nitroxide radicals:

$$R_1 = R_2 = (CH_2)_3CH_3 \text{ for trans-5-decene, and}$$
$$\begin{cases} R_1 = (CH_2)_7CH_3 \text{ or } (CH_2)_7COOH \\ R_2 = (CH_2)_7COOH \text{ or } (CH_2)_7CH_3 \text{ for oleic acid;} \end{cases}$$

(1)

and similarly the reaction with octadecene would yield the species:

(2)   (3)   (4)

where $R = (CH_2)_{15}CH_3$ for octadecene.

In all such nitroxide structures there have to be at least two hydrogens bonded to carbon atoms adjacent to the free radical center—i.e., β protons. One of the difficulties that we have encountered in assigning Type II radicals is that the electron spin resonance spectra for this species produced in either oleic acid or trans-5-decene gives no indication of hyperfine interaction other than to a nitrogen nucleus, while the corresponding radical produced on octadecene gives rise to a similar nitrogen hyperfine interaction with a further weak coupling of approximately $\sim$2.4 gauss (Fig. 10). McRae and Symons (1966) have reported, however, that nitroxide radicals produced by the reaction of NO_2 with methyl methacrylate give rise to β hydrogen hyperfine couplings of $\sim$11 gauss.

An alternative assignment of Type II radicals would be that they are nitrosoanion radicals. Such radicals are known to give rise to characteristic electron spin resonance spectra, with nitrogen hyperfine coupling constants of *ca* 13 gauss. They are also known to be stabilized by addition of bases such as triethylamine but to decay rapidly in acidic solution. Both basic solutions prepared by addition of triethylamine and acidic solutions prepared by addition of both acetic and hydrochloric acids were shown to have no observable effect upon the intensity of the electron spin resonance spectra of the hyperfine structure of Type II radicals produced from both oleic acid and octadecene. We, therefore, conclude that the stability of the species to acidic solution and the total absence of the conjugate acid form of the nitrosoanion radical must be regarded as strong evidence that such species are not present.

A further possibility is that the Type II radicals arise from a dimerization of nitroso compounds produced in the reaction of nitrogen dioxide with the unsaturated compounds. Theilacker and Uffmann (1965) have recently reported

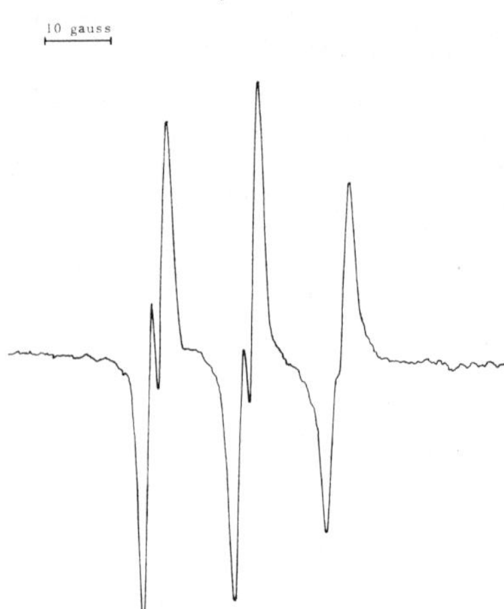

FIG. 10. ESR spectrum showing hyperfine structure of Type II radical from octadecene-NO₂-chloroform system.

observing electron spin resonance spectra from solutions of 2-methyl 2-nitroso-propyl acetate, which they have attributed to the formation of a dimeric species according to the following reaction scheme:

$$2R-N \rightleftharpoons R-\overset{\cdot}{N}-\overset{\cdot}{N}-R$$

(5)

They assume that for steric reasons, this species behaves as a biradical rather than as a species with a triplet ground state in that there exists no observable interaction between the two unpaired electrons. Such a species would, in fact, give rise to the nitrogen hyperfine coupling constants of $a_N = \sim 15$ gauss characteristic of a nitroxide radical. The possibility that such species are formed during the NO_2 reaction with both octadecene and trans-5-decene must be considered, since in each case the resulting solutions exhibit absorption maxima at 6600 Å characteristic of the presence of nitroso compounds. Oleic acid solutions, on the other hand, do not give rise to the characteristic 6600 Å absorption. The formation of such biradicals would, in the case of octadecene, lead to the following structures:

(6)   or   (7)

Structure 6 can be rationalized with the experimental observation of a basic triplet due to the coupling of an unpaired electron with one nitrogen nucleus

68

(coupling constant $a_N = 13.15$ gauss), together with a further weak splitting of each component into a doublet indicating hyperfine interaction with one proton with a coupling constant of a_H of approximately 2.4 gauss. Structure 7, however, cannot contribute to the Type II species since, based purely upon the well-known $B \cos^2 \theta$ relationship for the hyperfine coupling of β hydrogens, it is not possible for one of the methylene hydrogens to exhibit such a small hyperfine interaction and for the second to be non-observable.

From the available evidence, we must conclude that the Type II radicals in all systems are nitroxide radicals, but that their mode of formation remains somewhat inconclusive. If they are formed in the manner proposed by McRae and Symons (1966), then the absence of β hydrogen interactions must be fortuitous in that in each case the β hydrogens lie in the nodal plane of the nitrogen $2p$ orbital containing the unpaired electron. For this to be possible, structures 3 and 4 must either be not formed or have too short a lifetime to have been detected under the experimental conditions. Optical studies of oleic acid–nitrogen dioxide solutions lend support to the participation of the McRae and Symons (1966) mechanism in the formation of the nitroxide radicals since these solutions gave rise to no observable absorption at 6500 Å characteristic of the presence of nitroso species which are precursors of the nitroxide radicals in the Thielacker formulation.

For both octadecene and trans-5-decene, however, the reaction mixtures gave rise to the characteristic 6500 Å absorption indicating the presence of nitroso species, and the dimerization reaction proposed by Thielacker *et al.* (1965), appears to be favored.

Type III free radicals have been shown to be present in systems which contained appreciable concentrations of the Type II free radicals. They exhibit nitrogen hyperfine coupling constants and g factors characteristic of iminoxy radicals of general formula $RR_1\!-\!C\!=\!N\!-\!O$, to which they can be assigned.

In view of the findings that the formation of both Type II and Type III radicals, i.e., nitroxide and iminoxy radicals, are enhanced by ultraviolet radiation, it appears likely that the photochemical dissociation of NO_2 into nitric oxide and oxygen atoms plays an important role in their formation. Whether or not this dissociation reaction is a necessary precursor to the formation of these species will require detailed study. With the evidence available, the reaction mechanisms appear to be consistent with the following scheme:

$$RCH\!=\!CHR + NO_2 \longrightarrow R\overset{\cdot}{C}HCHNO_2R \longrightarrow RCH_2\overset{\cdot}{C}CH_2R \quad (1)$$

$$\text{initial} \qquad \underset{\underset{O \diagup \ \diagdown O}{N}}{}$$

Type I

$$R\overset{\cdot}{C}HCHNO_2R + NO_2 \longrightarrow RCHNO_2\text{-}CHNO_2R \qquad (2)$$

$$NO_2 \underset{}{\overset{h\nu}{\rightleftharpoons}} NO + O \qquad (3)$$

$$\text{RCH}{=}\text{CHR} + \text{NO} + \text{O} \longrightarrow \underset{\underset{\displaystyle \text{N}{=}\text{O}}{|}}{\text{RCH}}{-}\underset{\underset{\displaystyle \text{O}\cdot}{|}}{\text{CHR}} \qquad (4a)$$

$$\text{RCH}{=}\text{CHR} + \text{NO}_2 \longrightarrow \underset{\underset{\displaystyle \text{NO}_2}{|}}{\text{RCH}}{-}\text{CHR} \longrightarrow \text{RCH}{-}\underset{\underset{\displaystyle \text{O}\cdot}{|}}{\text{CHR}} \qquad (4b)$$

$$\underset{\underset{\displaystyle \text{N}{=}\text{O}}{|}}{\text{RCH}}{-}\underset{\underset{\displaystyle \text{O}\cdot}{|}}{\text{CHR}} + \text{RCH}{=}\text{CHR} \longrightarrow \overset{\text{RCH}{-}\text{CHR}}{\underset{\text{RCH}{-}\text{CHR}}{\text{O}{-}\text{N} \qquad \text{O}}} \qquad \text{Type II} \qquad (5a)$$

$$2\,\text{RCH}{=}\text{CHR} + \text{NO}_2 \longrightarrow \overset{\text{RCH}{-}\text{CHR}}{\underset{\text{RCH}{-}\text{CHR}}{\text{O}{-}\text{N} \qquad \text{O}}} \qquad \text{Type II} \qquad (5b)$$

$$\underset{\underset{\displaystyle \text{N}{=}\text{O}}{|}}{\text{RCH}}{-}\underset{\underset{\displaystyle \text{O}\cdot}{|}}{\text{CHR}} + \text{NO} \longrightarrow \text{RCH}{-}\overset{\displaystyle \text{ONO}}{\text{CHR}} \qquad (6)$$

$$2\,\underset{\underset{\displaystyle \text{N}{=}\text{O}}{\text{C}}}{\overset{\displaystyle \text{H}}{|}}\,\text{CHR} \longrightarrow \text{R}{-}\text{C}{-}\text{C}{-}\text{N}{-}\text{N}{-}\text{C}{-}\text{C}{-}\text{R} \qquad (7)$$

$$\text{Type II}$$

$$\text{R}{-}\underset{\underset{\displaystyle \text{N}{=}\text{O}}{|}}{\text{CH}}{-}\underset{\underset{\displaystyle \text{O}\cdot}{|}}{\text{CHR}} \longrightarrow \text{R}{-}\overset{\displaystyle \text{OH}}{\underset{\underset{\displaystyle \text{N}{=}\text{O}}{\|}}{\text{C}}}{-}\text{CHR} \qquad \text{Type III} \qquad (8)$$

Reactions 5a and 5b appear to predominate in the formation of the nitroxide radicals observed in the oleic acid reaction, whereas the available evidence favors reaction 7 for both octadecene and trans-5-decene.

The formation of nitrite derivatives was indicated by the absorption system extending from *ca* 3800 to 3350 Å in all reactions studied. Reaction 6 is suggested as a possible reaction leading to the formation of such derivatives, consistent with the available data. We have so far no direct evidence of the presence of the oxy radical intermediate.

REFERENCES

Arioka, I. (1967). *Hokkaido Igaku Zasshi* **40**, (9–12), 457.
Felmeister, A., Amanat, M., and Weiner, N. D. (1968). *Environ. Sci. Technol.* **2**, 1.
Gross, J. M., and Symons, M. C. R. (1966). *J. Chem. Soc.* (A), 451.

Levy, N., and Rose, J. D. (1947). *Quart. Rev. (London)* **1**, 358.
McRae, J. A., and Symons, M. C. R. (1966). *Nature* **206**, 1259.
Rieger, P. H., and Fraenkel, G. K. (1963). *J. Chem. Phys.* **39**, 609.
Rowlands, J. R., Estefan, R. M., Gause, E. M., and Montalvo, D. A. (1968). *Environ. Res.* **2**, 47–71.
Symon, K. (1965). *Proc. Roy. Soc. Med.* 988.
Theilacker, W., and Uffman, H. (1965). *Angew. Chem. Intern. Ed.* **4**, 8.
Thomas, H. V., Mueller, P. K., and Wright, R. (1967). *J. Air Pollution Control Assoc.* **17**, 33.
Thomas, H. V., Mueller, P. K., and Lyman, R. L. (1968). *Science* **159**, 532.

Carbon Monoxide as Pollutant and
Its Analytic Detection

Continuous Determination of Carbon Monoxide by Frontal Analysis

L. Dubois and J. L. Monkman

THE MEASUREMENT of the concentration of carbon monoxide in uncontaminated air or in samples taken in remote areas is difficult to achieve because of low sensitivity of the methods presently available.

In a previous paper, we presented preliminary studies on the use of frontal analysis for the sampling and measurement of monoxide in air (*1*). As was then demonstrated, this new method of sampling has tremendous advantages over all the other methods of sampling using adsorbents.

EXPERIMENTAL

An instrument was built at that time to make use of this principle of frontal analysis. To further improve the measurement of carbon monoxide, another instrument, which will be described, has been built. At the moment, the apparatus permits the continuous determination of carbon monoxide concentrations in air for a minimum of a six-minute period. Figure 1 illustrates the apparatus schematically; it consists mainly of a system of four columns

---

(1) L. Dubois and J. L. Monkman, *Mikrochim. Acta*, 2, 313–320 (1970).

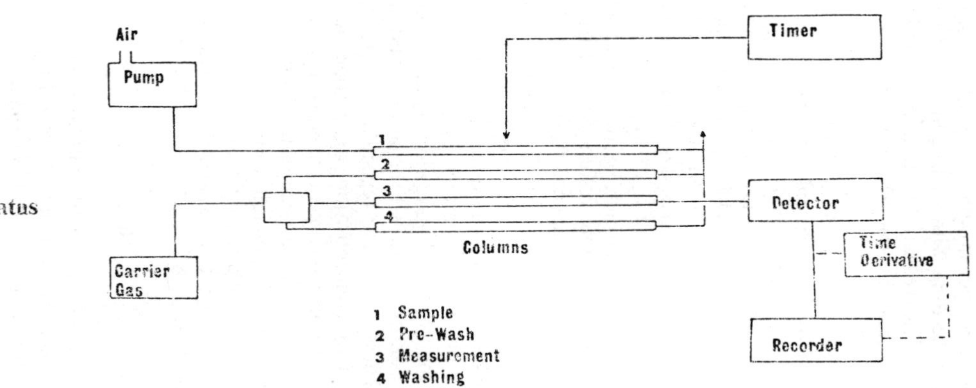

Figure 1. Diagram of the apparatus

in which different operations are taking place simultaneously. On the first column, sampling is carried out by passing air at a flow rate of 75 ml per minute. On the second column, pre-washing is proceeding with hydrogen at a flow rate of 10 ml per minute. The reason for this pre-washing is to elute nitrogen and oxygen, which have been sampled on the adsorbent, so that they are not passed through the flame detector. They will extinguish the flame if they are allowed to go through. On the third column, the CO measurement is being made by elution with H_2 at a flow rate of 30 ml per minute. The eluted gases are directed to a nickel catalyst heated at 250 °C where CO is reduced to methane (2). A hydrocarbon detector, Beckman 109, is used to measure the converted carbon monoxide. The fourth column is washed or regenerated by a flow of hydrogen at 200 ml per minute. The purpose of this treatment is to remove any compound having a higher retention volume than carbon monoxide. Columns 1, 2, and 4 are opened to the atmosphere. With a system of 16 solenoid valves located at the inlet of the column and four located at the outlet operated by a timer, it is possible to change, after any desired period of time, 2, 3, 4, 5, 10, etc. minutes, to another cycle where sampling will be done on the fourth column, washing on the third, measurement of carbon monoxide on the second, and pre-washing on the first. These cycles are repeated automatically and continuously. The concentration of carbon monoxide measured, will be some kind of an average over a 2, 3, 4, 5, or 10-minute period. By decreasing the length of the cycle to a minimum time, it is possible to get closer and closer to an instantaneous measurement of the carbon monoxide concentration. Practically, it is difficult to go to a cycle shorter than five minutes.

RESULTS

Figure 2, shows the results obtained with a 10-ppm CO standard in air, the instrument operating on a six-minute cycle. The sensitivity control on the Beckman 109 hydrocarbon analyzer was set at 30× and a 5-mV span voltage Philips recorder was used. The chromatogram shows where the change from one column to the other takes place. The peak which appears at the beginning of the cycle is due to an increase in the pressure of the system during the automatic switching of the solenoid valves. The height of the plateau, above the base line, is proportional to the concentration of carbon monoxide. Good reproducibility can be obtained with the four different columns which must have similar

---

(2) K. Porter and D. H. Volman, ANAL. CHEM., **34**, 748 (1962).

properties, that is, same length, same amount of adsorbent, same retention time, etc. On the extreme left, an air sample containing 1 ppm carbon monoxide has been measured.

At the top of Figure 3, the instrument has been used in a ten-minute cycle so that the complete carbon monoxide elution chromatogram could be shown. This time, a 5-ppm carbon monoxide standard in air was used and again the attenuator of the Beckman 109 hydrocarbon analyzer was set at 30, and the same 5-mV span voltage recorder was used. Once more, good reproducibility is obtained. At the end of the four cycles, the signal of the Beckman 109 hydrocarbon analyzer was fed to a Cahn time derivative computer, the output of which was in turn fed to a 1-mV span voltage recorder. The bottom part of Figure 3 was then obtained. The chief reason for taking the first derivative is to eliminate problems which could occur in measurement of the height of the plateau due to drifting of the base line. Another area where the first derivative could be useful might be in case one would like to use this technique to measure more than one component. In this situation a two, three, or four-plateau chromatogram will result; the use of the first derivative will then obviate continuous switching of the sensitivity control to keep the signal on scale.

The height of the peak obtained by the first derivative should be also proportional to the concentration of carbon monoxide. One drawback to the use of the first derivative is a decrease in sensitivity as shown in Figure 3. With the usual method, concentrations of carbon monoxide as low as 0.1 ppm are easily measured, and this sensitivity is then decreased to 2 ppm if the time derivative computer is introduced into the

Table I.   Carbon Monoxide Concentrations in Laboratory Air

Time	CO, ppm
10:00	0.22
10:06	0.24
10:12	0.24
10:18	0.28
10:24	0.32
10:30	0.31
10:36	0.30
10:42	0.31
10:48	0.31
10:54	0.46
11:00	0.57

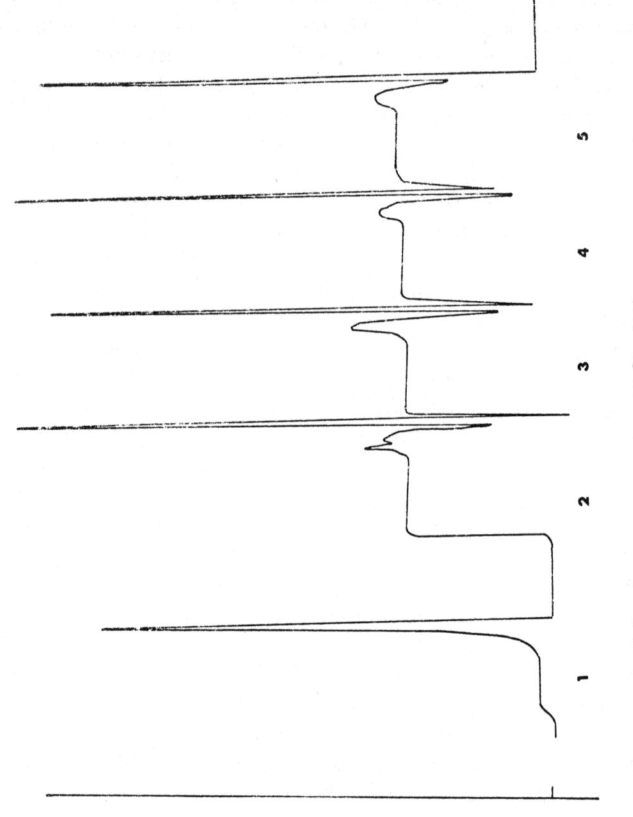

Figure 2. Chromatogram obtained with a
CO standard and an air sample

78

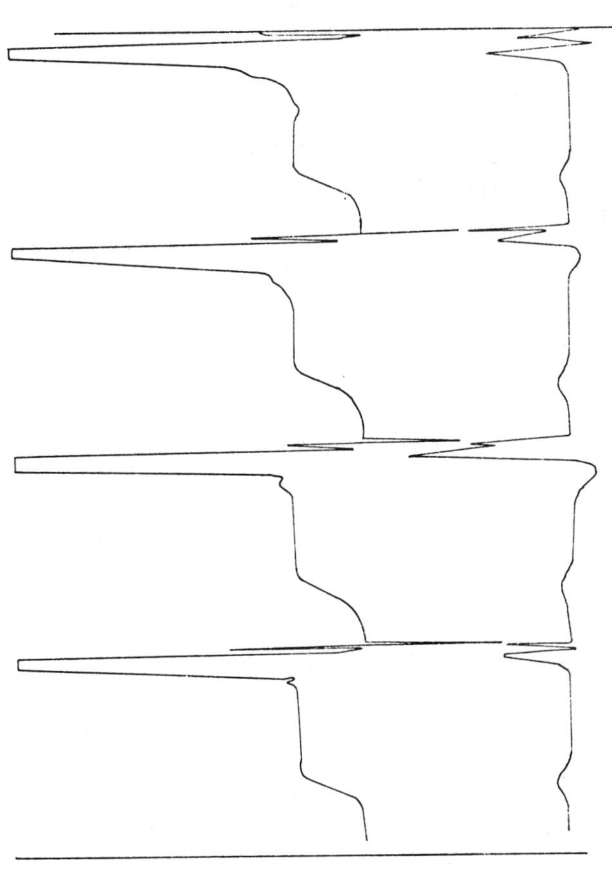

Figure 3. Chromatograms obtained with a CO standard in the usual mode (upper curves) and in the first derivative mode (lower curves)

system. This technical problem should, however, be easily solved.

In Table I are given results obtained for carbon monoxide concentrations in laboratory air with the instrument operating without the time derivative attachment. These results are typical of the measurements which can be made with this apparatus.

DISCUSSION

From the results obtained to date with this instrument, it seems feasible to measure on a routine basis the concentration of carbon monoxide in uncontaminated air and in air samples taken at base-line stations of the World Meteorological Organization (WMO). In the near future, we should be able to evaluate the minimum amount of carbon monoxide which can be measured accurately with this technique. From the chromatograms obtained on air samples, it seems possible that methane could also be measured at the same time as carbon monoxide. Further investigations of these possibilities are proceeding.

ACKNOWLEDGMENT

We thank Howard Chung for the art work.

CONTINUOUS MEASUREMENT OF CARBON MONOXIDE IN STREETS 1967–1969

L. E. REED and P. E. TROTT

Abstract—Continuous measurements of carbon monoxide were made at a number of sites in towns in Britain for periods of 13–15 months. High concentrations were not found at any of the sites. The proportion of time each month when carbon monoxide concentration exceeded 30 ppm ranged from below 0·01 to 1·92 per cent.

INTRODUCTION

THE ROLE the motor vehicle has played in polluting the atmosphere in California has been established beyond doubt and concern about similar occurrences in this country has been often expressed. The smog which occurs in California is however mainly caused by interaction between the hydrocarbons emitted from motor vehicles and photochemically produced oxidants. Although such a situation seems unlikely in Northern Europe because of differences in climate and land configuration, the emission in the streets of large volumes of carbon monoxide from petrol engined cars has received some attention. Instantaneous measurements (CHOVIN, 1967) of carbon monoxide concentrations in Paris streets over several years have shown that about 10 per cent of them exceeded the Threshold Limit Value (MINISTRY OF LABOUR, 1966, AMERICAN CONFERENCE OF GOVERNMENTAL INDUSTRIAL HYGIENISTS, 1968) of 50 ppm recommended in Britain for factory atmospheres, although admittedly this limit is the time-weighted average for a normal working day. In view of this and the fact that increasing numbers of petrol engined vehicles are being licensed to operate on British roads there is a need for more information on the concentration of carbon monoxide in streets.

This report describes continuous measurements made at one site in each of several cities in Great Britain during 1967–1969. The survey was carried out on behalf of the Ministry of Transport to obtain information on concentration levels of carbon monoxide found in the street and the frequency and duration of their occurrence.

All concentrations in this report are quoted as parts per million by volume because this is the unit most commonly used both in this country and abroad.

EXPERIMENTAL

Site description

The first observations were made at Luton from June 1966 to July 1967. The main survey was then carried out between September 1967 and March 1969 at Glasgow, Birmingham, Manchester, Enfield, Portsmouth and Cardiff, each site being in operation for 13–15 months.

All sites were in central urban streets, well known locally as carrying a high volume of traffic and close to traffic lights or pedestrian crossings which interfered with the traffic flow (see Appendix).

81

Sampling lines and instruments

The inlets to the instruments were at a height of 2·75 m (9 ft) vertically above the edge of the kerb. The sampling lines were of 5 mm bore ($\frac{1}{4}$ in. o.d.) brass tubing (except in Glasgow for which 6 mm bore p.v.c. tubing was used) and their lengths were between 7 and 30 m (25 and 100 ft) the inlet end of each sampling line was fixed to a lamp-post or other post and fitted with a small downward facing polythene funnel the sole purpose of which was to prevent drops of water from being sucked up the line. Between the sampling line and the instrument were two drying towers filled with magnesium perchlorate and a filter tube containing animal wool. At the sampling rate of 0·5–0·75 l. min^{-1} the air from the street required 2–3 min to reach the instrument, most of the delay occurring in the drying towers.

The measuring instruments were all non-dispersive i.r. gas analysers. They were calibrated five or six times per week at two points on the scale. Zero was set on air which had passed through a hopcalite train to remove carbon monoxide. The second calibration point was at about three quarters of full scale deflection and was checked by introducing a standard gas mixture of the appropriate concentration (150 ppm CO in most cases). Each i.r. gas analyser was connected to a recording millivoltmeter which also operated a series of counters to indicate the length of time during which the concentration of carbon monoxide exceeded certain levels. Counter readings were normally taken five or six times per week. The counters at Luton were set to record times above 20, above 40, above 60 ppm and several higher values. At the other sites one counter was set for above 50 ppm because since 1966 this has been the Threshold Limit Value recommended by the Factory Inspectorate, MINISTRY OF LABOUR (1966). Another counter was set for above 10 ppm this being considered the maximum reading likely to arise from background carbon monoxide and error due to zero drift on the analyser. The remaining counter was set midway at above 30 ppm which was in fact the "Serious Level for 8 hours" Standard for the STATE OF CALIFORNIA (1959).

Since these measurements were completed, new standards have been proposed for the STATE OF CALIFORNIA (1969). The recommended air quality standard for CO is now 20 ppm for an 8 h period.

The i.r. gas analysers and recorders were operated on a scale capable of registering peaks of up to 180 ppm but could have been switched to a scale with full scale deflection 450 ppm had it become necessary. An instrument with a more sensitive scale would have been useful for lower concentrations but of course would not have registered the high peaks.

In order to check on errors which might arise between visits (often 72 h apart at weekends) the CO concentration was noted every night for the period 01.00–06.00 h, at which time traffic flow is very low. On most nights the recorder charts indicated a concentration of 0 ± 5 ppm.

The variations of recorded CO concentration from site to site particularly at higher values, could be caused by differences in concentration of the calibration gas mixtures (nominally 150 ppm) used at each site. To assess this error the instruments were checked during February–March 1968 against a single gas cylinder transported to each site in turn. This mixture of nominal concentration 130 ppm was analysed on each instrument immediately after calibration on its individual gas mixture. The spread of values was 135–149 ppm which was very small compared with the variations in observed CO concentration.

All instruments were run continuously throughout the test periods except for minor breakdowns. The periods lasted for at least 13 months except in the case of Cardiff where before a continuous electricity supply to the instrument became available in June, it only operated between approximately 07.30 and 18.00 h from Mondays to Saturdays.

Very little time was lost by the instruments being out of action or giving unreliable results. From the number of days on which measurements were acceptable (that is all days shown in TABLE 1), it can be estimated that, taking all sites together, the instruments were operating satisfactorily for 93 per cent of the time they were installed.

RESULTS

The length of time above a given concentration for each month is shown in TABLE 1. These times have also been presented as percentages of the total time of operation except when this amounted to less than one week per month or during March–June 1968 at Cardiff.

TABLE 1. DURATION OF CO CONCENTRATION ABOVE A GIVEN VALUE

Site	Duration of measurements		Minutes above			Per cent above		
	In days	In minutes	10 ppm	30 ppm	50 ppm	10 ppm	30 ppm	50 ppm
20 Sept–2 Oct 1967								
Birmingham	12	16,929	200·4	0·1	0	1·18	<0·01	0
2 Oct–1 Nov 1967								
Birmingham	30	42,838	351·9	4·6	0·2	0·82	0·01	<0·01
Manchester	15	21,456	3302·1	16·2	0·4	15·39	0·08	<0·01
Enfield	8	11,588	178·0	2·0	0	1·54	0·02	0
Glasgow	1	1429	168·0	2·3	1·3	—	—	—
1 Nov–1 Dec 1967								
Birmingham	30	41,308	1523·3	7·9	0·2	3·69	0·02	<0·01
Manchester	30	42,966	13,754·8	751·9	43·2	32·01	1·75	0·10
Enfield	30	42,966	3308·0	95·0	2·0	7·70	0·22	<0·01
Glasgow	26	36,976	5587·0	84·0	8·4	15·11	0·23	0·02
1 Dec 1967–2 Jan 1968								
Birmingham	32	45,873	544·8	4·1	0·5	1·19	<0·01	<0·01
Manchester	32	45,761	10,554·0	429·4	30·2	23·06	0·94	0·07
Enfield	31	44,603	2727·6	149·6	2·9	6·12	0·34	<0·01
Glasgow	31	44,502	7749·0	99·1	7·5	7·41	0·22	0·02
2 Jan–2 Feb 1968								
Birmingham	31	44,197	259·5	1·0	0·5	0·59	<0·01	<0·01
Manchester	30	42,890	9335·9	213·3	4·7	21·77	0·50	0·01
Enfield	31	44,430	772·0	24·4	0	1·74	0·05	0
Glasgow	30	43,005	11,210·0	76·8	3·5	26·07	0·18	<0·01
2 Feb–1 Mar 1968								
Birmingham	28	39,911	1265·1	10·3	0·8	3·17	0·03	<0·01
Manchester	28	40,047	8641·3	219·5	7·7	21·58	0·55	0·02
Enfield	28	40,224	3604·3	70·1	2·0	8·96	0·17	<0·01
Glasgow	14	20,285	3385·0	27·3	2·7	16·69	0·13	0·01
Cardiff	3*	1426	246·4	0·5	0·2	—	—	—
Portsmouth	7	10,007	299·2	0·8	0	2·99	<0·01	0

TABLE 1. (cont.)

Site	Duration of measurements		Minutes above			Per cent above		
	In days	In minutes	10 ppm	30 ppm	50 ppm	10 ppm	30 ppm	50 ppm
1 Mar–2 Apr 1968								
Birmingham	30	42,218	330·8	0·3	0	0·78	<0·01	0
Manchester	31	44,235	5254·0	80·6	3·6	11·88	0·18	<0·01
Enfield	32	45,937	820·0	13·0	0·5	1·79	0·03	<0·01
Glasgow	27	38,776	4022·0	17·1	0·9	10·37	0·04	<0·01
Cardiff	26*	15,006	2350·1	140·5	5·9	—	—	—
Portsmouth	27	38,681	5970·4	4·2	3·4	15·44	0·01	<0·01
2 Apr–2 May 1968								
Birmingham	28	40,446	562·0	4·5	0·3	1·39	0·01	<0·01
Manchester	27	38,579	3891·1	69·8	3·0	10·09	0·18	<0·01
Enfield	30	42,976	850·4	26·0	0	1·98	0·06	0
Glasgow	28	40,490	11,155·0	182·2	4·1	27·55	0·45	0·01
Cardiff	23*	13,025	64·3	7·5	4·2	—	—	—
Portsmouth	30	42,945	4851·8	1·6	1·1	11·30	<0·01	<0·01
2 May–31 May 1968								
Birmingham	29	41,501	1280·9	11·5	0·3	3·09	0·03	<0·01
Manchester	25	35,532	6751·3	110·4	2·8	19·00	0·31	<0·01
Enfield	29	41,914	3099·2	100·8	7·1	7·39	0·24	0·02
Glasgow	28	40,253	16,271·0	52·2	1·3	40·42	0·13	<0·01
Cardiff	28*	17,642	69·4	7·8	4·6	—	—	—
Portsmouth	18	25,540	1808·7	1·2	0	7·08	<0·01	0
31 May–1 July 1968								
Birmingham	24	34,170	383·1	4·7	1·0	1·12	0·01	<0·01
Manchester	31	44,546	1166·6	7·0	0·2	2·62	0·02	<0·01
Enfield	31	44,053	1901·9	67·3	1·6	4·32	0·15	<0·01
Glasgow	31	44,353	14,024·0	15·1	0·8	31·62	0·03	<0·01
Cardiff	30*	34,110	146·4	17·4	4·7	—	—	—
Portsmouth	24	34,817	14,919·1	16·7	0·1	42·85	0·05	<0·01
1 July–1 Aug 1968								
Birmingham	31	44,252	858·4	4·6	1·2	1·94	0·01	<0·01
Manchester	31	44,520	7857·0	106·7	2·9	17·65	0·24	<0·01
Enfield	31	44,489	3087·1	80·7	3·1	6·94	0·18	<0·01
Glasgow	23	32,517	2865·0	31·9	1·8	8·81	0·10	<0·01
Cardiff	29	41,673	7803·5	89·9	1·5	18·73	0·22	<0·01
Portsmouth	30	42,963	8065·8	11·9	0·2	18·77	0·03	<0·01
1 Aug–30 Aug 1968								
Birmingham	27	38,610	292·0	1·7	0	0·76	<0·01	0
Manchester	29	41,315	6481·4	44·3	0·8	15·69	0·11	<0·01
Enfield	29	41,568	2370·7	46·4	1·5	5·70	0·11	<0·01
Glasgow	29	41,955	4159·0	16·9	0	9·91	0·04	0
Cardiff	29	41,653	8723·6	144·5	1·2	20·94	0·35	<0·01
Portsmouth	29	41,518	13,662·8	3·2	0	32·91	<0·01	0
30 Aug–3 Oct 1968								
Birmingham	34	48,540	1196·5	6·4	0·7	2·47	0·01	<0·01
Manchester	31	44,359	6710·7	154·5	4·8	15·13	0·35	0·01
Enfield	34	49,278	2584·3	119·3	2·6	5·24	0·24	<0·01
Glasgow	31	43,962	5498·0	34·2	2·7	12·51	0·08	<0·01
Cardiff	33	47,373	7931·7	248·3	6·5	16·74	0·52	0·01
Portsmouth	—	—	—	—	—	—	—	—
3 Oct–1 Nov 1968								
Birmingham	29	41,263	2193·1	13·2	0·3	5·31	0·03	<0·01
Manchester	24	34,501	7434·6	325·4	13·0	21·55	0·94	0·04
Enfield	28	40,400	4383·2	93·6	3·4	10·85	0·23	<0·01
Glasgow	29	41,906	3830·0	22·1	0·1	9·14	0·05	<0·01

84

TABLE 1. (cont.)

Site	Duration of measurements		Minutes above			Per cent above		
	In days	In minutes	10 ppm	30 ppm	50 ppm	10 ppm	30 ppm	50 ppm
3 Oct–1 Nov 1968 (*cont.*)								
Cardiff	26	37,372	6583·0	321·4	38·9	17·61	0·86	0·10
Portsmouth	29	41,369	12,824·5	39·5	1·1	31·00	0·10	<0·01
1 Nov–3 Dec 1968								
Birmingham	32	45,827	3274·7	28·4	2·8	7·15	0·06	<0·01
Manchester	32	45,684	4422·8	157·5	9·7	9·68	0·34	0·02
Enfield	32	45,617	5508·4	294·6	15·6	12·08	0·65	0·03
Glasgow	30	42,965	3715·0	19·7	1·6	8·65	0·05	<0·01
Cardiff	27	38,776	5186·6	416·8	70·1	13·38	1·07	0·18
Portsmouth	32	46,032	3582·0	13·4	0·1	7·78	0·03	<0·01
3 Dec 1968–3 Jan 1969								
Birmingham	31	44,608	2347·7	18·5	0·5	5·26	0·04	<0·01
Manchester	29	42,782	4752·7	163·2	12·3	11·11	0·38	0·03
Enfield	31	44,390	6707·1	324·0	18·5	15·11	0·73	0·04
Glasgow	28	40,260	9326·0	247·1	8·8	23·16	0·61	0·02
Cardiff	26	37,341	6946·4	715·2	63·0	18·60	1·92	0·17
Portsmouth	31	44,470	970·9	11·9	0·5	2·18	0·03	<0·01
3 Jan–3 Feb 1969								
Cardiff	29	41,764	2523·3	202·3	25·3	6·04	0·48	0·06
Portsmouth	31	44,432	917·2	4·7	1·1	2·07	0·01	<0·01
3 Feb–3 Mar 1969								
Cardiff	28	40,100	3809·2	233·4	10·5	9·50	0·58	0·03
Portsmouth	10	14,068	3·5	0·1	0	0·02	<0·01	0
3 Mar–31 Mar 1969								
Cardiff	28	40,197	4944·3	188·1	15·7	12·30	0·47	0·04
Portsmouth	28	40,187	636·5	13·2	1·0	1·58	0·03	<0·01
Total								
20 Sept 1967–31 Mar 1969								
Birmingham	458	652,491	16,864·2	121·8	9·3	2·58	0·02	<0·01
Manchester	425	609,173	100,310·3	2849·7	139·3	16·47	0·47	0·02
Enfield	435	624,433	41,902·2	1506·8	60·8	6·71	0·24	<0·01
Glasgow	386	553,634	90,442·0	1046·1	45·5	16·34	0·19	<0·01
Cardiff†	255	366,247	54,451·6	2759·9	232·7	14·87	0·75	0·06
Portsmouth	326	467,029	68,512·4	122·4	38·6	14·67	0·03	<0·01

* In operation for 10·5 h only on most days.
† All days with less than 24 h sampling omitted.

The results obtained in 1966–1967 at Luton were assessed at different concentration levels. During 536,966 min of operation, 432 min (0·08 per cent of total) were above 20 ppm and 16 min above 40 ppm.

Although the time above 50 ppm amounted to only a few minutes each month this might possibly include some much higher concentrations. The recorder charts were therefore examined to find the value and frequency of the peak concentrations. TABLE 2 shows the number of occasions when the concentration exceeded 50 ppm and TABLE 3 the maximum concentration recorded each period. Electrical interference also caused sudden movement of the recorder pen and it was difficult, in some cases, to distinguish genuine from spurious values.

85

TABLE 2. FREQUENCY OF PEAKS ABOVE 50 ppm

Period	Sites					
	Birming-ham	Man-chester	Enfield	Glasgow	Cardiff	Ports-mouth
1 Nov–1 Dec 1967	3	29	3	2	—	—
1 Dec 1967–2 Jan 1968	1	41	8	2	—	—
2 Jan–2 Feb 1968	1	6	1	0	—	—
2 Feb–1 Mar	1	17	3	0*	—	0*
1 Mar–2 Apr	0	6	2	0	3*	0
2 Apr–2 May	1	5	1	5	0*	0
2 May–31 May	0	7	11	2	0*	0
31 May–1 July	3	1	5	1	2*	0
1 July–1 Aug	2	12	7	4	3	1
1 Aug–30 Aug	0	4	4	0	2	0
30 Aug–3 Oct	2	10	8	5	8	—
3 Oct–1 Nov	2	24	7	0	33	2
1 Nov–3 Dec	4	21	26	4	55	0
3 Dec 1968–3 Jan 1969	2	24	27	11	44	1
3 Jan–3 Feb 1969	—	—	—	—	20	1
3 Feb–3 Mar	—	—	—	—	20	0*
3 Mar–31 Mar	—	—	—	—	15	1

* Results obtained during periods of less than 2 weeks or from days of less than 24 h.

TABLE 3. HIGHEST PEAK CONCENTRATION ON CHART, ppm

Period	Sites					
	Birming-ham	Man-chester	En-field	Glas-gow	Car-diff	Ports-mouth
1 Nov–1 Dec 1967	68	82	58	53	—	—
1 Dec 1967–2 Jan 1968	55	84	57	57	—	—
2 Jan–2 Feb 1968	62	75	51	43	—	—
2 Feb–1 Mar	67	69	84	42*	—	30*
1 Mar–2 Apr	31	64	53	45	54*	30
2 Apr–2 May	53	75	50	60	30*	32
2 May–31 May	49	66	81	76	31*	36
31 May–1 July	60	55	70	56	65*	46
1 July–1 Aug	66	70	112	60	61	51
1 Aug–30 Aug	37	53	71	48	53	36
30 Aug–3 Oct	60	74	63	83	79	—
3 Oct–1 Nov	53	80	86	46	87	73
1 Nov–3 Dec	60	70	114	55	92	49
3 Dec 1968–3 Jan 1969	56	98	87	93	107	62
3 Jan–3 Feb 1969	—	—	—	—	83	72
3 Feb–3 Mar	—	—	—	—	83	17*
3 Mar–31 Mar	—	—	—	—	81	53

Period	Date of peak	Birming-ham	Man-chester	En-field	Glas-gow	Car-diff	Ports-mouth
1 Nov 1967–3 Jan 1969	3.11.67	68					
	31.10.68						73
	30.11.68			114			
	14.12.68		98				
	28.12.68				93		
	2.1.69					107	

* Results obtained during periods of less than 2 weeks or from days of less than 24 h.

86

Diurnal variation

Examination of the recorder charts showed that CO concentrations were generally below 10 ppm between 24.00 (midnight) and 08.00 h on weekdays and all day on Sundays. The diurnal variation occurring on weekdays is illustrated by FIGS. 1 and 2 which show 15 min average concentrations derived from the recorder chart trace on a Monday and a Wednesday in the same week at the same site. These concentrations are fairly typical of days of high concentration (FIG. 1) and of low concentration (FIG. 2) both at this site and the others. For example there is a definite period of higher concentration during the evening rush hour traffic period at 16.00–19.00 h and a less pronounced higher concentration between 08.00 and 10.00 h.

When the peaks above 50 ppm shown in TABLE 2 were classified into hours of the day, they were found to occur most frequently during the early evening. Three quarters of the peaks at Cardiff, half of those at Birmingham, Manchester, and Enfield and one quarter of those at Glasgow occurred during the 3-h period 16.00–19.00 h. The hour between 17.00 and 18.00 h was by far the worst of the day at Cardiff, Manchester and Enfield. No hour outside the early evening period had an unusually high proportion of peaks at any of the sites.

Proportion of time spent at higher concentrations

The number of minutes above 10, 30 and 50 ppm were noted at each site for the periods between successive readings of the counters. These were normally 24 h during the week and longer at the weekends. TABLE 1 shows the sums of these numbers of minutes for each month, the start of a month being chosen as a day on which all the sites were visited.

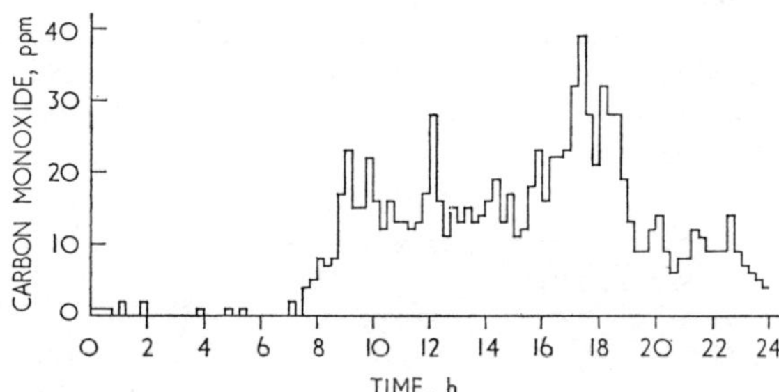

FIG. 1. Fifteen minute mean concentrations on a weekday, Manchester, Monday 4.11.68.

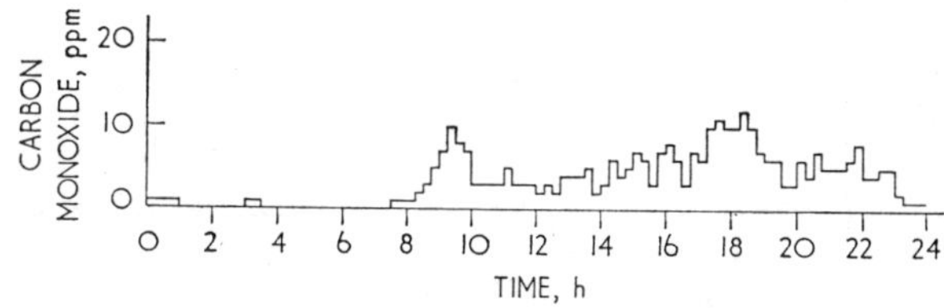

FIG. 2. Fifteen minute mean concentrations on a weekday, Manchester, Wednesday 6.11.68.

The percentages of total time spent above 30 ppm shown at the end of TABLE 1 are less than 1 per cent at all sites. This is in agreement with earlier measurements elsewhere. For example GEORGII and WEBER (1962) found that in Frankfurt during 1959–1961, 99 per cent of all values were below 30 ppm even at the point with the highest traffic density. CLAYTON *et al.* (1960) found that over a period of 58 weeks in 1956–1957 the CO concentration was in the range 0–100 ppm with a median of 10 ppm at the most polluted sampling site in Detroit. In central London during 1956–1958 LAWTHER *et al.* (1962) measured concentrations exceeding 200 ppm by instantaneous sampling and 10 min mean values exceeding 50 ppm by continuous sampling. During 1961–1962 by continuous sampling throughout 5 months they found similar CO levels with a mean concentration of 23 ppm for the hour of 17.00–18.00 h on Mondays–Fridays and a mean of 7·8 ppm for the whole of the time (LAWTHER *et al.* 1962; WALLER *et al.* 1965).

Peak concentrations

Although for the present measurements, sampling was continuous, there was a delay of 2–3 min between the air entering the inlet and the signal reaching the recorder. Sharp fluctuations in concentration as individual vehicles passed the inlet were thus partly smoothed out by mixing on the way to the instrument. Most of the delay occurred in the drying towers, some of it in the sampling tube from the inlet and less than 0·5 min in the measuring cell of the instrument itself. FIGURE 3 shows the concentrations recorded at a flow rate of 0·75 l. min^{-1} when pure air was replaced by air containing 150 ppm CO for a period of 6·5 min. The recorder reading reached 90 per cent of the true reading after 2 min 30 s. If CO is injected for a period shorter than the delay time, the peak recorded on the instrument will never reach the full value. FIGURE 4 shows for example that injection of 150 ppm CO for 1 min produced a maximum reading of 121 ppm.

The figures for peak concentrations in TABLES 2 and 3 are divided into the same periods as were used in TABLE 1 but commencing 1st November 1967 by which time four out of the six sites were in operation. TABLE 3 also gives the date of the highest peak found at each site.

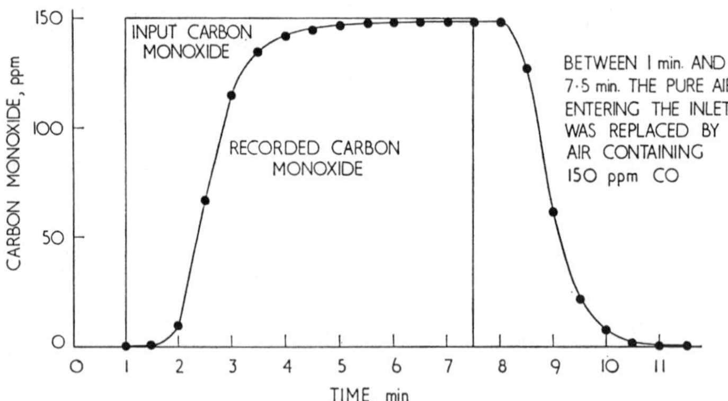

FIG. 3. Response of analyser.

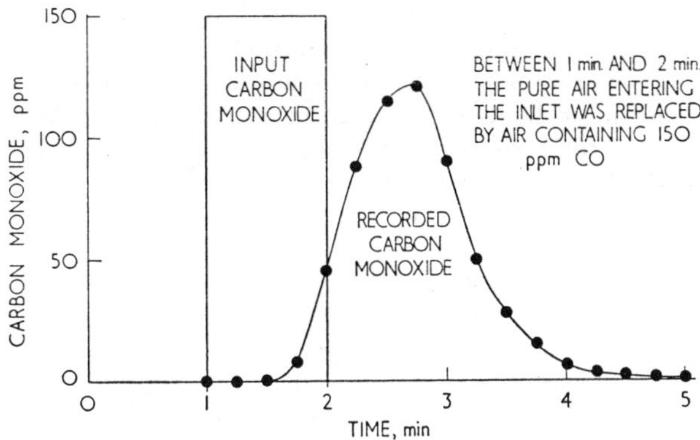

FIG. 4. Response of analyser to a 1 min dose.

No peak concentrations reached 120 ppm although higher peaks have been reported on several occasions in the past. These were in most cases found in instantaneous samples collected in plastic bags over 5–15 s, which would occasionally coincide with short high peak concentrations. In Paris, CHOVIN (1967) found that the "real average risk" to which a human being would be exposed is 3–4 times less than the maximal risk as found from instantaneous samples. His "real average risk" was evaluated by continuously recording the CO concentration and integrating the records to obtain 15 min averages.

The duration of peak concentrations above the 50 ppm level was investigated for comparison with the Threshold Limit Value of 50 ppm averaged over 8 h inside factories. An examination was made of all single days or groups of days (in cases when daily readings were not taken) occurring during the 20 months in TABLE 1 when 5 min or more was spent above the 50 ppm level. Only the following three days had more than 15 min above 50 ppm.

<pre>
17–18 November 1967, (1 day), Manchester, 15·6 min
17–18 October 1968,   (1 day), Cardiff,   19·9 min
18–19 November 1968, (1 day), Cardiff,   28·1 min.
</pre>

The maximum concentrations on these days were 78, 82 and 86 ppm respectively.

Examination of the recorder charts showed that concentrations above 50 ppm mostly occurred in sharp peaks of short duration. The individual peaks with the longest times above 50 ppm ever found in this survey were six lasting 5 min and one lasting 8 min.

Even allowing for the smoothing of sharp peaks because of the delay time in sampling, the duration of peak concentrations above the 50 ppm level must be far below the Threshold Limit Value (AMERICAN CONFERENCE OF GOVERNMENTAL INDUSTRIAL HYGIENISTS, 1968). However this is a standard recommended for healthy people in factories who presumably spend the rest of the 24 h in less polluted surroundings. The limit for outside air should obviously be concerned with the effect on the general population of exposure to varying concentrations over the whole 24 h and the possibility that some individuals may be particularly sensitive to CO must be borne in mind.

The STATE OF CALIFORNIA (1959) Department of Health considered that since 10 per cent carboxyhaemoglobin in the blood would produce a significant effect on oxygen transport by the blood and since half of this could be derived from other sources such as cigarette smoking or faulty stoves, then the ambient air should not give rise to more than 5 per cent carboxyhaemoglobin. Calculations using the equations of FORBES *et al.* (1945) showed that 5 per cent would normally be reached after 8 h exposure at 30 ppm or 1 h at 120 ppm by an individual engaged on light work. These exposures were therefore termed the Serious Level in California.

Since these measurements were completed, the recommended standard for the STATE OF CALIFORNIA (1969) has become 20 ppm CO for an 8 h period and will probably be 50 ppm for 1 h exposure. Such concentrations can produce an increase of over 2 per cent in carboxyhaemoglobin and there is some evidence that this increase may impair important Central Nervous System functions. At the same time the report says that moderate cigarette smoking produces an average carboxyhaemoglobin concentration substantially greater than 2 per cent.

The CO concentration in the blood is always tending to reach equilibrium with the CO in the air, as has been illustrated (CHOVIN, 1967) by testing the blood of traffic policemen before and after duty. Non-smokers showed an increase in blood CO, whereas smokers who started work with high carboxyhaemoglobin levels tended to exhale CO. This means that a number of peaks at intervals during a day would be expected to be less dangerous than one continuous period at the same concentration.

The duration of peak concentrations above the 30 ppm level was investigated by examining the times above 30 ppm for every single day or group of days (in cases when daily readings were not taken). Only 25 of these were found to have more than 1 h above 30 ppm, the four with the longest times being,

18–20 November 1967,	(2 days),	Manchester,	239·3 min
21–22 October 1968,	(1 day),	Manchester,	120·4 min
29 November–2 December 1968,	(3 days),	Cardiff	203·8 min
27–29 December 1968,	(2 days),	Glasgow,	159·4 min.

The time above 30 ppm on these four occasions was not continuous.

The "real average risk" of CHOVIN (1967) referred to earlier was based on 15 min averages. The possibility of a 15 min average above 30 ppm will obviously only arise on days with at least 15 min total above 30 ppm. As there were only 148 such days or groups of days out of the total of 2329 days considered between 1st November 1967 and 31st March 1969, the number of days which include a 15 min period with an average concentration above 30 ppm must be small.

Effect of sampling height

The inlet to the sampling tube was fixed at 2·75 m (9 ft) above ground level so as to be fairly inaccessible to passers by and yet be not far removed from the breathing area of people in the street. Measurements in Paris (CHOVIN, 1967) have been made at 1·7 m (5 ft 6 in.) the nose height of an average man at points in the centre of the road. Such positions are not normally available in other countries because of the difficulty of sampling in the roadway without disturbing the traffic flow. The measurements in Fleet Street, London reported by LAWTHER *et al.* (1962) and WALLER *et al.* (1965) were made at a height of 1·2 m (4 ft) from a pipe on the railings of a public lavatory. Sites such as this one, on islands in the middle of the roadway, are unfortunately rare and do not always coincide with high traffic density. The CO concentrations recorded by the Continuous Air Monitoring Program (JUTZE and TABOR, 1963) in major cities in the U.S.A. were lower than at the roadside. At their sites the sampling inlet is 3·7–4·6 m (12–15 ft) above ground level on a building usually situated some distance away from the traffic.

GEORGII (1966) found that the vertical distribution of CO in a street depended on the meteorological conditions and whether the sampling point was on the exposed windward side of a building or whether it was on the opposite side of the street in the lee and protected from the wind. For a measuring position on the leeward side he found a gradient of 10 ppm 30 m^{-1} (that is a concentration of 18 ppm at 3 m falling to 8 ppm 33 m^{-1} above ground level) during calm conditions. This gradient reduced to 8 ppm at 30 m when the wind increased to 8 m s^{-1}, whereas at the exposed windward position windspeeds above 3 m s^{-1} were sufficient to produce a uniform concentration of 5 ppm independent of height. Our own measurements of vertical distribution at Luton on a day when the wind speed was 2·2 m s^{-1} (5 m.p.h.) did not disagree with this but were inconclusive owing to the low concentration of CO and have not been reported.

Seasonal variation

In order to compare summer and winter figures the average proportions of time above given CO concentrations in minutes per day have been calculated from TABLE 1. These proportions which appear as TABLE 4 refer to the year 2nd February 1968 to

TABLE 4. AVERAGE TIME PER DAY ABOVE A GIVEN CO CONCENTRATION

CO concentration (ppm)	Site	Min day^{-1}		
		Winter	Summer	Year (2.2.68–3.2.69)
> 10		62·7	26·4	43·3
> 30	Birmingham	0·5	0·2	0·3
> 50		0·03	0·02	0·02
> 10		211·8	188·8	199·8
> 30	Manchester	6·6	2·8	4·5
> 50		0·32	0·08	0·19
> 10		139·2	75·5	104·2
> 30	Enfield	5·3	2·4	3·7
> 50		0·25	0·09	0·17
> 10		189·7	288·1	245·8
> 30	Glasgow	3·5	2·0	2·6
> 50		0·11	0·06	0·08
> 10		174·0	143·8	157·2
> 30	Cardiff	13·1	3·0	7·5
> 50		1·48	0·13	0·73
> 10		156·5	330·6	235·7
> 30	Portsmouth	0·5	0·3	0·4
> 50		0·04	0·01	0·03

The last column refers to average minutes per day for the period 2nd February 1968 to 3rd February 1969, summer is the 6 months April to September 1968.

3rd February 1969 with the 6 months April–September 1968 being taken as summer. The proportions of time above 30 and above 50 ppm were greater in winter than in summer but it should be emphasized that this is based on one year only and that there was considerable variation between sites.

CONCLUSIONS

During the whole period of measurements in 1967–1969 the proportion of time each month when 30 ppm was exceeded ranged from below 0·01–1·92 per cent. Normally only a few minutes of this time were above 50 ppm, the maximum during one month at any site being 70 min, and the longest single peak above 50 ppm only lasting 8 min. The highest peak concentration recorded throughout this survey was 114 ppm.

Acknowledgements—We are very grateful to the staff of the Public Health Departments of Luton, Glasgow, Manchester, Enfield, Cardiff and Portsmouth and the Department of Building of the University of Aston for their assistance in choosing the sites and in visiting the instruments five or six times each week. We are also indebted to the owners of the premises mentioned in the Appendix for allowing us to install the instruments.

REFERENCES

AMERICAN CONFERENCE OF GOVERNMENTAL INDUSTRIAL HYGIENISTS (1968) *Threshold Limit Values for 1968*. The Conference Secretary–Treasurer, Cincinnati, Ohio.

CHOVIN P. (1967) Carbon monoxide: Analysis of exhaust gas investigations in Paris. *Environ. Res.* **1**, 198–216.

CLAYTON G. D., COOK W. A. and FREDERICK W. G. (1960) A study of the relationship of street level carbon monoxide concentrations to traffic accidents. *J. Am. Ind. Hyg. Ass.* **21**, 46–54.

FORBES W. H., SARGENT F. and ROUGHTON F. J. W. (1945) The rate of carbon monoxide uptake by normal men. *Am. J. Physiol.* **143**, 594–608.

GEORGII H.-W. and WEBER E. (1962) Untersuchung der Kohlenoxyd-Immission in einer Grossstadt (Investigation of the ground level concentration of carbon monoxide in a large town). *Int. J. Air Water Pollut.* **6**, 179–195.

GEORGII H.-W. (1966) Die Vertikalverteilung des Kohlenmonoxid in Grossstadtstrassen in Abhängigkeit von den Meteorologischen Bedingungen. (The vertical distribution of carbon monoxide in city streets related to meteorological conditions.) Paper VI/18. *Proceedings of International Clean Air Congress, October* 1966. The National Society for Clean Air, London.

JUTZE G. A. and TABOR E. C. (1963) The continuous air monitoring program. *J. Air Pollut. Control Ass.* **13**, 278–280.

LAWTHER P. J., COMMINS B. T. and HENDERSON M. (1962) Carbon monoxide in town air: An interim report. *Ann. Occup. Hyg.* **5**, 241–248.

MINISTRY OF LABOUR (1966) *Factory Inspectorate, Safety Health and Welfare Booklet, New Series No. 8, Dust and Fumes in Factory Atmospheres*, 3rd edn, H.M. Stationery Office, London.

STATE OF CALIFORNIA (1959) Dept. Public Health, *Technical Report of California Standards for Ambient Air Quality and Motor Vehicle Exhaust*, pp. 14–15 and 69–73. The Department, Berkeley, California.

STATE OF CALIFORNIA (1969) Air Resources Board, *Recommended Ambient Air Quality Standards*. State Department of Public Health, Berkeley, California, May 21, 1969.

WALLER R. E., COMMINS B. T. and LAWTHER P. J. (1965) Air pollution in a city street. *Br. J. Ind. Med.* **22**, 128–138.

APPENDIX—DESCRIPTIONS OF SITES

Luton

Eastern Gas Board Showrooms, 68 George Street, Luton, Beds. In operation June, 1966 to August, 1967. The inlet was on a traffic light standard at the NW end of the main shopping street. At this point the main street forks to pass on either side of the Town Hall. Traffic passes one-way round the Town Hall. The site was on the NE side of the street which means that traffic proceeding towards the SE used the lane next to the sample inlet. The instrument was operated by the Public Health Department, County Borough of Luton.

Glasgow

Glasgow Corporation Transport Offices, 46 Bath Street, Glasgow, C.2. First installed November 1966 but the time counters started on 31st October, 1967. Measurements ceased on 31st December, 1968. The inlet was on a kerbside post not far from the traffic lights at the corner of Renfield Street. Bath Street is an important route for westbound traffic through the central area and Renfield Street is an important street for southbound traffic. Both are one-way streets wide enough for four lanes of traffic. The site was on the N side of the street which means that the right-hand lane of westbound traffic was next to the sample inlet. The instrument was operated by the Public Health Department, City of Glasgow. This site was under the supervision of Warren Spring Laboratory, Scottish Branch.

Birmingham

Police Station, Digbeth, Birmingham, 5. Started 20th September, 1967. The inlet was on a post about 20 m before traffic lights which control traffic proceeding towards the ESE. The site was only about 0·25 mile from the Bull Ring on the road called Digbeth which is the main traffic route between the city centre and Coventry and London. There are three lanes of traffic in each direction separated by a narrow central strip. The site was on the NNE side of the street. The instrument was operated by the Department of Building, University of Aston.

Manchester

Corporation Offices, 80, Mosley Street, Manchester 2. In operation 13th October, 1967 to 4th January, 1969. The site was in the centre of the city at a crossroads junction, both streets having one-way traffic. The inlet was on the traffic light standard on the NE side of Princess Street next to the left hand lane of traffic and only about 5 m from the left-hand lane of traffic proceeding towards the NE along Mosley Street. Princess Street is four lanes wide and Mosley Street is three lanes wide. The instrument was operated by the Health Department, Corporation of Manchester.

Enfield

G.P.O., Church Street, Enfield, Middlesex. In operation 20th October, 1967 to 3rd January, 1969. The inlet was on a lamp-post about 2 m beyond the pedestrian (Zebra) crossing outside the Post Office in the main shopping street. There was one lane of traffic in each direction and since the site was on the N side of the street the eastbound lane was the one nearest to the inlet. Besides local traffic there is some heavy traffic between the M.1 and the Thames-side docks. The instrument was operated by the Health Department, London Borough of Enfield.

Cardiff

Woodhouse's, 47, Queen Street, Cardiff. In operation 27th February, 1968 to 31st March, 1969. The inlet was on a traffic light standard on the N. side of Queen Street which is the main route for west bound traffic through the city centre. The street carries four lanes of one-way traffic, the site being next to the right hand lane at a light controlled pedestrian crossing which is just before a side street on the left. The instrument was operated by the Public Health Department, City of Cardiff.

Portsmouth

Landport Drapery Bazaar, Commercial Road, Portsmouth, Hants. In operation 22nd February, 1968 to 31st March, 1969. The inlet was on a lamp standard outside a large store on the east side of Commercial Road in the centre of the city. The road is wide enough for two lanes of traffic in each direction. Traffic from the north is held up about 3 m beyond the sampling point by a policeman on point duty at the junction with Edinburgh Road. The instrument was operated by the Public Health Department, City of Portsmouth.

Levels of Carbon Monoxide Recorded on Aircraft Flight Decks

H. J. JUDD

CARBON MONOXIDE IS A GAS which is toxic to man when inhaled in abnormal concentrations. Carbon monoxide is produced in many ways but our main interest was concentrated with that produced in the exhaust fumes of aircraft. Danger can arise in aircraft if the concentration of carbon monoxide on the flight deck exceeds a certain level. Hitherto a maximum level of 50 ppm was thought to be acceptable since this figure is an empirical one used in industrial premises. However, as a result of our investigations we believe 30 ppm should be the maximum, with respect of aircraft and aircrew.

This danger is aggravated by present traffic density at airports where circumstances, all too common, delay aircraft for long periods before take-off. The problem was highlighted when the crew of a Boeing 707 aircraft was forced to queue for 2 hours and 50 minutes at John F. Kennedy Airport before take-off. The crew reported a feeling of nausea and dizziness which may have been caused by an excess of carbon monoxide. In view of this reported incident, Dr. Bennett, Chief Medical Officer of the Board of Trade, approached Dr. F. S. Preston, Principal Medical Officer (Air), with regard to carrying out a series of tests to measure the levels of carbon monoxide in queueing aircraft.

TOXICOLOGY OF CARBON MONOXIDE

Carbon monoxide is a colorless, tasteless, odorless gas which results from the incomplete combustion of organic matter, i.e. petrol, kerosene, diesel oil. The toxic effects of carbon monoxide are due to its action on the blood, whereby it produces a deficiency of oxygen in the tissues. Haemoglobin, the pigment present in the red blood cells, normally combines with oxygen in the lungs and carries it to the body tissues. Unfortunately, haemoglobin is not selective in the types of gas with which it will combine and will take up carbon monoxide in the same way as oxygen, forming carboxyhemoglobin instead of oxyhemoglobin; in fact its affinity to carbon monoxide is 210 times greater than for oxygen. Thus, when carbon monoxide competes with oxygen for a place in the hemoglobin molecule, the odds are 210 in favor of carbon monoxide to the exclusion of oxygen. Both gases cannot be carried by the same molecule at the same time. Therefore, a very small concentration of carbon monoxide can inactivate a large amount of hemoglobin as an oxygen carrier. In addition, the remaining active hemoglobin is affected in such a way that it binds its oxygen more tenaciously and does not give it up readily to the body tissues. As a result, a state of oxygen deficiency is produced which is identical to the oxygen deficiency produced at high altitude.

McFarland[2] has shown by experiment that light sensitivity of the eye is the most delicate functional test for oxygen deficiency and has shown that as little as 3% carboxyhemoglobin has produced measurable impairment. Furthermore, the effects of carbon monoxide and altitude are cumulative. The effect of a given increase in carbon monoxide saturation was roughly the same as that of an equal loss of arterial oxygen saturation due to high altitude. On this basis one can determine the "Physiological Altitude" that results in the blood after smoking.

As already stated, carbon monoxide is one of the products of incomplete combustion of any organic matter, including tobacco. Approximately 1-2.5% of total volume of cigarette smoke is carbon monoxide. About half the carbon monoxide present in the inspired air is absorbed, amounting to 10-15 cu.cm. per cigarette. The total gas-containing capacity of the hemoglobin of an average man is about 1,000 cc. Thus, smoking one cigarette results in the saturation of 1-1.5% of the body's hemoglobin with carbon monoxide. It has been shown that an average smoker will have approximately 5.7% carbon monoxide saturation of the hemoglobin, and a person smoking 20-30 cigarettes a day could be as high as 10%.

Figure 1, shows the combined effects of altitude and carbon monoxide in amounts that may result from cigarette smoking, i.e. if a pilot flying at 10,000 ft. true altitude has a 10% saturation of his blood with carbon monoxide, he would experience a physiological effect equivalent to 15,000 ft. Point "A". McFarland.[3]

When the exposure to carbon monoxide is terminated dissociation from the hemoglobin occurs, but the average time required for the percentage of carboxyhemoglobin in the blood to drop to half its value is approximately 250 minutes. The process could naturally be speeded up by the inhalation of pure oxygen; under these conditions dissociation will occur in approximately 40 minutes.

There are, however, circumstances which are extremely relevant when considering the effects of carbon monoxide, namely length of time of exposure, cardiac output, the oxygen demand of the tissue and the hemoglobin content of the blood, i.e. anemic individuals are more susceptible and increased metabolic rate increases the severity of the symptoms.

Table I shows the relationship between carboxyhemoglobin and symptoms from severe to acute, but warning symptoms, however, may not occur. In these circumstances muscular weakness leading to collapse and coma are the initial onset. When a high concentration of carbon monoxide is suddenly inhaled, transient weakness and dizziness may be the only premonitory warning before unconsciousness; even these may be lacking sometimes. What makes the problem more acute is the differing thresholds of the individuals.

As already mentioned, an anemic individual is more susceptible to carbon monoxide, and heavy smokers al-

ready have a percentage of carboxyhemoglobin in their
blood. Thus it is possible for circumstances to conspire
adversely and a situation occur whereby an aircrew on
arrival at the airport may already have been exposed

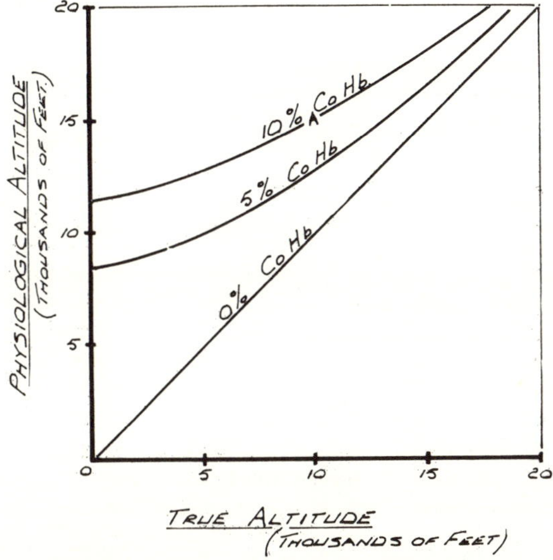

Fig. 1. Effects of carbon monoxide on altitude tolerance.

to exhaust fumes and industrial atmospheric pollution
(they may have been caught up in a serious traffic jam)
and should such a crew consist of heavy smokers, it
will be appreciated that such a group of individuals
will be more susceptible to carbon monoxide through
previous exposure to the gas (Table II).

Table II indicates the levels of carbon monoxide
measured inside a motor vehicle travelling within and
outside the city limits of Washington, D.C. It will be
appreciated that the levels of atmospheric pollution in
areas such as Los Angeles, New York and San Francisco would contain a higher level than Washington,
D.C. Although it is known that small amounts of carbon monoxide affect visual acuity, it is not known
whether the accepted maximum of 50 ppm causes any
mental impairment. For these reasons it seems highly
desirable that further work should be done as recommended.

TABLE I. SYMPTOMS

Symptoms	% of Blood Saturation
No symptoms	0–10
Tightness across forehead—slight headache—dilation of cutaneous blood vessels	10–20
Headache—throbbing in temples	20–30
Severe headache—weakness—dizziness—dimness of vision—nausea—vomiting—collapse	30–40
Same as before but more severe and increased respiration and pulse	40–50
Syncope—increased respiration and pulse—coma with intermittent convulsions—Cheyne—Stokes respiration	50–60
Coma as above—depressed heart action and respiration—possible death	60–70
Weak pulse and slowed respiration respiratory failure—death	70–80

TABLE II. IN TRAFFIC CARBON MONOXIDE DATA—WASHINGTON.
15TH-27TH APRIL, 1967.[4]

Route	Time Minutes Average Running	Miles Distance	Runs No. of	Co. PPM Average	Co. PPM Max 5 Min
Arterial	30	9.1	17	31	56
Expressway	36	17.4	17	34	66
City Centre	15	2.2	34	45	93

TABLE III. EMISSIONS OF AIR CONTAMINANTS FROM COMMER-
CIAL AIRCRAFT OPERATING AT LOS ANGELES AIRPORT, (1)
(THESE FIGURES CAN BE EQUATED TO MOVEMENTS AT
LONDON HEATHROW AIRPORT).

Year	Air Traffic at Lax. Flights/Day	Fuel Consumption Gallons Per Day	Particulate Matter	CO	NO_2	Hydro Carbons	SO_2
1969	973	213,000	7	16	5	40	2
LHR. estimated increase for 1975, based on average increase 1964/69	1,400	378,000	10.5	24	7.5	60	3

METHODOLOGY

After extensive investigations, through a variety of instrument manufacturers, it was decided that the only effective portable instrument that could be located on the flight decks of aircraft was the National Coal Board type (T.C.O.III). The instrument is based on measuring the amount of heat evolved when carbon monoxide is oxidized to carbon dioxide.

$$CO + \tfrac{1}{2}O_2 \longrightarrow CO_2 = 68 \text{ K Cals} \qquad (283 \text{ KJ})$$

The catalyst used is a granular mixture of manganese dioxide and copper oxide and the reaction occurs at ambient temperature. Inserted into the catalyst are two thermistors which act as differential thermometers and measure the temperature rise in the catalyst caused by the oxidation of the carbon monoxide. The instrument was calibrated daily with a standard gas/air mixture and its accuracy is $\pm$ 0.001% in the 0-300 ppm range. The instrument was arranged both in position and time to record the levels of pollution on the flight deck which was being breathed by the aircrew.

METEOROLOGY

The atmosphere disperses air contaminants. The rate at which dispersion occurs will depend on speed, direction and variability of air movement. In general any atmospheric condition which causes a layer of still air will increase the degree of contamination within the layers. Still air layers may occur at any altitude depending on the conditions. The rate of vertical dispersion is generally determined by the character of temperature distribution with height. If the stable layer is based at the ground, pollutants tend to remain near the level at which they are emitted and to diverge slowly and horizontally. If the stable layer is aloft, pollutants accumulate at its base and at day-break, when the rising sun warms the air beneath, inversion occurs, bringing the pollutants to ground level where they may remain for an hour or so. Buildings, topographical features, cloudiness, sunshine and sea-land configuration all affect air movements and therefore dispersion rate. Thus, winds which will generally disperse contamination may, paradoxically, due to aerodynamic downwash, cause areas of high contamination and may also bring into an area pollution from industrial plants some considerable distance away. Nevertheless, area-wide contamination levels consistently decrease as wind speed increases. The relation between wind direction and pollution sources determine the pattern of contamination. The time of year is relevant since there is more atmospheric stagnation in the winter months with the resulting buildup of general atmospheric pollution. Rain will partially wash out pollution, while snow will scour it out.

Meteorological conditions are of the utmost importance in a study of this type. They cannot be ignored for it is possible that all adverse factors could be present at one time. Unfortunately, no extreme inversions occurred during this study.

PRESSURIZATION

Aircraft require cabin pressurization and air conditioning to compensate for the variation in barometric pressure and in temperature at the altitude at which they operate. The cabin pressurization system maintains the cabin at a lower altitude than that at which the aircraft is flying, while the air conditioning system maintains a comfortable cabin air temperature.

At 35,000 ft. the barometric pressure is 3.5 psi and the air temperature −65°F. So, to enable an acceptable cabin altitude of 6,500 ft. (barometric pressure 11.5 psi) to be achieved, the cabin must be pressurized to a cabin/aircraft altitude differential pressure of 8.0 psi. The air required for this purpose is generally provided by engine driven compressors or by bleed air from the engine compressor, e.g. on the Boeing 707-436 aircraft the total air supply from two engine bleeds and the two compressors is approximately 3,000 cubic feet/minute at 300°F.

Before this air enters the cabin it passes through a temperature control valve and, if necessary, is routed through air conditioning packs to provide the desired temperature.

Cabin pressure and the rate of change of cabin altitude is regulated by a controller set to the desired conditions. (The normal cabin altitude change is 300 ft/minute). Signals from the controller operate valves opening to atmosphere which regulate the cabin pressure by bleeding excess air overboard. In the event of a pressurization failure, oxygen masks for passengers automatically drop from their containers when the cabin altitude reaches 13,500 ft.

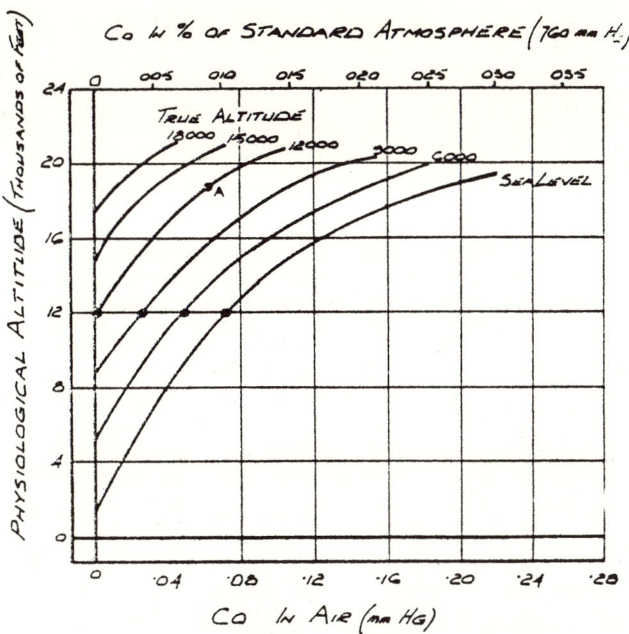

Fig. 2. The relation between physiological altitude and the partial pressure of carbon monoxide in air at various true altitudes when equilibrium with blood has been reached.

It must be appreciated that partial pressurization to 0.26 in. of mercury is implemented at the time the hatches are closed and therefore air, which could be heavily contaminated from the exhausts of leading aircraft, internal combustion engines, ground handling equipment and industrial effluents, will be drawn in through the engines and then circulated throughout the aircraft interior during taxiing and at holding position.

Pressure has a direct influence on the effect of carboxyhemoglobin. McFarland[2] has shown that a carboxyhemoglobin saturation of 16% at an altitude of 9,000 ft. produces a "physiological" altitude of 16,000 ft. The author shows that a concentration of only 0.08% carbon monoxide at an altitude of 9,000 ft. could produce these results. (Figure 2).

DISCUSSION

A difficulty which presented itself during the study of the levels of carbon monoxide on the flight deck was the absence of meteorological variables, and throughout the exercise no meteorological conditions, inversions, fog, or low stratus which could cause retention at ground levels of exhaust gas, were met. Consequently, there was no aircraft queueing or delay.

TABLE IV. MEASUREMENTS FROM 13TH OCTOBER 1969 TO 15TH DECEMBER 1969 AT LONDON—EUROPEAN—JOHN F. KENNEDY AIRPORTS.

This table is a summary of the recordings, using a type T.C.O. III Carbon Monoxide Detector and Recorder kindly loaned from the National Coal Board. It will be seen that in spite of some quite moderate wind speeds, appreciable carbon monoxide levels were recorded.

Airport	Aircraft Type	Takeoff & Direction	Wind Speed (KTS) & Direction	Time Instrument Switched "ON" to take off mins	Max. Co levels recorded ppm	
LHR	Trident	100°	7/160°	47	4	
LHR	Trident	280°	10/200°	31	—	
FRA	Trident	070°	4/082°	45	5	
LHR	Trident	280°	12/270°	46	10	
ZRH	Trident	280°	5/250°	48	12	
LHR	Trident	280°	9/250°	40	8	
LHR	Trident	100°	5/110°	44	6	
LHR	Trident	280°	12/250°	35	5	
MAD	Trident	330°	8/300°	51	7	
LHR	Trident	280°	8/165°	31	16	
LIS	Trident	180°	12/150°	34	7	
LHR	Trident	280°	2/170°	32	5	MINIMUM 15 MINUTE PERIOD
LHR	Trident	280°	5/250°	43	—	
LHR	Trident	280°	5/235°	29	20	
LHR	Trident	280°	1/260°	39	15	
DUS	Trident	180°	10/160°	48	5	
LHR	Trident	280°	5/230°	36	15	
LHR	B.707	280°	11/240°	46	24	
BDA	VC10	090°	10/100°	44	—	
JFK	B 707	040°	8/050°	77	15	
JFK	VC 10	310°	10/360°	49	20	
BDA	VC 10	090°	5/105°	64	—	
JFK	B 707	310°	15/340°	73	36	
JFK	B 707	220°	5/200°	70	5	
JFK	B 707	220°	14/285°	39	22	
JFK	DC 9	310°	10/330°	41	8	
YUL	DC 9	100°	3/110°	59	5	
JFK	B 707	310°	14/340°	72	10	

Indications of organic combustion efflux are shown on Table III. They are the results of a study carried out at Los Angeles Airport and equate almost identically to the number of movements at London Heathrow (Table IV). However, from the data obtained from some 30 observations, the levels of carbon monoxide did not

102

reach any serious proportions, i.e. at no time did the
levels reach 50 ppm which was the accepted maximum
for an eight-hour exposure.

As this study was of a preliminary nature it is considered that the survey (in which aircrews and cabin staff
may be monitored) should be continued to (a) determine the average level of carboxyhemoglobin present
in air crew before boarding aircraft, (b) determine what
level of carboxyhemoglobin can be accepted for a "GO"
state and (c) determine if the agreed standard level of
carboxyhemoglobin is likely to be exceeded during taxiing and holding due to the ingestion of carbon monoxide from the exhaust of leading aircraft, in which case
a strict "No Smoking" order, similar to that of alcohol,
should apply.

It is further considered that pilots and airport authorities should be warned of the hazards of queueing
for protracted periods in cases where following aircraft
can ingest fumes from leading aircraft. A holding pattern as shown in Figure 3 should be adopted. It should
also be determined whether the level of 50 ppm of carbon monoxide results in mental impairment.

The Air Corporations Joint Medical Service is, in fact,
taking part in an air pollution survey jointly with the

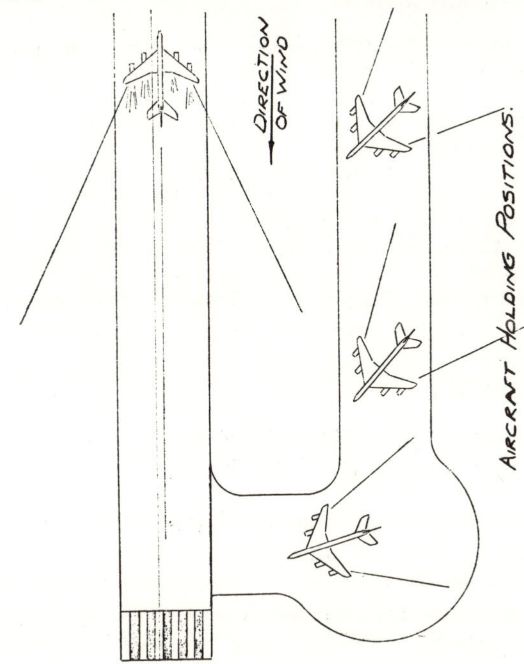

Fig. 3. Aircraft holding positions.

Board of Trade and the Ministry of Technology and has taken the initiative of bringing to the attention of the I.C.A.O. and I.A.T.A. Committees, respectively, a possible hazard to aircrews of atmospheric pollution and also the feasibility of installing monitoring points along taxi-ways at very busy airports. In this way information could be automatically fed back to Air Traffic Control so that additional control of aircraft movements along taxi-ways could be maintained, particularly when the levels of carbon monoxide were high.

Airport authorities and airlines should ensure that ground handling equipment using internal combustion engines be regularly and efficiently maintained and includes a catalytic unit on the exhaust of this type of equipment in order to oxidize the carbon monoxide to carbon dioxide.

Since publication of this paper, the British Airports Authority has informed airline operators using London Airports that petrol-driven ground handling equipment and vehicles will not be permitted after 1972 and all other internal combustion units will have to incorporate an efficient catalytic unit.

ACKNOWLEDGEMENTS

The author wishes to thank Dr. F. S. Preston, Principal Medical Officer (Air), Dr. D. Bruton, Principal Medical Officer (Ground), and Dr. James Munro, Consultant Anesthetist, Hillingdon Hospital, for their continued guidance and advice during this investigation.

Thanks are due to Mr. Wood, Chief Scientific Officer of the National Coal Board, North Western Region, Mr. Valentine, Chief Physicist of the National Coal Board, North Western Region, Captain Dell and Captain Meagher for their assistance and support.

I wish to thank Dr. J. Graham Taylor, Director Medical Services, A.C.J.M.S. for permission to publish this paper.

REFERENCES

1. LEMKE, E. E., G. THOMAS and W. E. SWAIKER: Profile of Air Pollution Control District, County of Los Angeles, 1969.

2. McFARLAND, R. A.: *J. Av. Med.* 15:28, 1944.

3. McFARLAND, R. A.: Human Factors in Air Transport. 174-175, 304-306.

4. Washington, D.C., Metrop. Area Air Poll. Abatement Activity, Nov. 1967. U.S. Dept. of Health, Education and Welfare.

Carbon Monoxide Research—Recent and Remote

John R. Goldsmith, MD

Several recent meetings and reports of current research suggest need to reevaluate the role of carbon monoxide exposures on the functions of the heart and central nervous system.

In a search of the literature for references to the possible effects of carbon monoxide on heart disease, a report of what was called "Shinshu myocardosis" was found. In the Japanese journal, *Digest of Science of Labour* (vol 10, pp 315-318, 1955), Dr. Fumio Komatsu reported on studies made in Kinasa village in the northern part of Shinshu district in the Nagano prefecture of Japan. The village had a population of slightly over 6,000 and is at an elevation of 700 to 1,000 meters above sea level, surrounded by mountain ranges. Because of the limited arable land area and cold climate, the economy depended on silk raising and lumber cutting in warm weather and the manufacture of tatami mats during the long winters. The indoor winter work was carried out in the kakoi (enclosed room) in which the windows and crevices in the walls are sealed. The temperature was kept customarily at 12 to 15 C, using an open charcoal fire; it was often crowded and provided an average of 5 cu m of air volume per person.

Komatsu reported that carbon dioxide levels were in the range of 1% to 3.4%, and the carbon monoxide was between 0.2% and 0.3% in the morning. He reported that the carboxyhemoglobin levels reach 20 to 30%. The workers often complained of headache and vertigo, called "sumiatari" (sickness from the charcoal fire) by the inhabitants. In a survey of the health status of the population, it was found that many persons were diagnosed as having valvular heart disease at city hospitals though they had no murmurs!

In the spring, after the long winter exposures, many of the subjects felt exhausted and had vertigo and facial edema. In the early winter and spring, there was a high frequency of angina pectoris-like attacks. Heart disease was the most common cause of death among village inhabitants. The rate was more than five times the average for the entire country. Komatsu performed physical examinations on 1,022 inhabitants; 35.3% had abnormal heart findings. Excluding obvious cases of hypertension, rheumatic and congenital heart disease, and other possible causes, 18% had a specific syndrome which Komatsu designated as Shinshu myocardosis. It occurred in people younger than was common with arteriosclerotic heart disease. There were even a few cases under age 20 years. The number of female patients was approximately double that of males.

Komatsu reported that the early stage of the condition is associated with stiffness of the shoulders, backache, and, occasionally, in response to interview questions, fatigue, vertigo, edema of the face while looking down during work. Electrocardiographic changes were observed even during this stage of the condition. During the second stage, subjects had dyspnea on exertion, substernal tightness, and pain at work, with some numbness of the upper extremities. Many of the symptoms were relieved by rest from work. During the next stage, the symptoms resembled those of paroxysmal nocturnal dyspnea. In a more severe stage, angina pectoris-like attacks occurred even with light work.

Under treatment, the condition persisted in some cases for as long as three to five years. Komatsu found cardiac enlargement by percussion and roentgenography, decreased muscular strength and vital capacity, and electrocardiographic changes including arrhythmia, low voltage, depressed S-T segments, and prolongation of the ventricular complex, especially the Q-T.

Komatsu stated that the Department of Hygiene at Shinshu University considered the main cause to be repeated inhalation of CO in small amounts over a long period of time. He reported that when the kakoi were remodeled in Kinasa village there was a decrease in the cardiac morbidity to one half. Changes in agriculture have now made the work in the kakoi unnecessary.

The article by Komatsu probably deserves recognition as the earliest systematic study of the relation between CO exposure and myocardial disease. The translation was prepared through the courtesy of Prof Toshio Toyama of the Department of Preventive Medicine and Public Health of Keio University Medical School.

The relevance of the paper is heightened by the recent meeting of the New York Academy of Sciences on Jan 12 to 14, 1970. At this meeting, two new facets of the relationship of CO to human health were reported and, in general, accepted as part of the scientific literature in this field. These are effects on the heart and central nervous system. The new emphasis on the heart was partly based on studies by Dr. Astrup and his colleagues in Denmark on the effect of CO in increasing deposition of lipids in the large blood vessels. The Astrup group also studied a group of smokers who had clinical peripheral and central atherosclerosis and who showed a substantially higher carboxyhemoglobin level than individuals with similar smoking histories without atherosclerosis. At the same meeting, both Drabkin and Preziosi reported on experimental work in animals in which myocardial damage was reported. Stupfel from France reported that low concentrations of CO (50 ppm) produced transient ventricular wave changes in the electrocardiogram (ARCH ENVIRON HEALTH 18:613, 1969). Ayres reported further studies on the effect of CO exposure in impairing myocardial metabolism, adding to the data presented earlier in the ARCHIVES 18:699, 1969.

In occupational exposures of firemen, Goldsmith stated that Gordon and Rogers observed high levels of carboxyhemoglobin (to 44%) from fighting fires of more than five minutes duration in Denver. Roughly one-half of the exposures were associated with electrocardiographic changes; often these were simply changes in rate or rhythm. There were some instances of increased atrioventricular conduction time and non-specific S-T wave changes.

At the New York Academy meeting, there was fairly extensive discussion of the effects of carbon monoxide on central nervous system function, which centered around Beard and Wertheim's report that exposures to 50 ppm for as little as 90 minutes led to significant decrements in time-interval estimation. A possible effect of CO exposure on estimation of distances between following automobiles was reported by Rockwell from Ohio State University. His exposures resulted in 5% and 10% carboxyhemoglobin levels, substantially above those likely to result on the basis of Beard's work. Two other groups headed by Mikulka and Hanks studied groups of experimental subjects, giving challenging tracking tests while exposed to CO. They observed no effects. Neither were effects found by Stewart's group at Marquette University. (this issue of the ARCHIVES, pp   ). All three of these latter groups exposed subjects under group conditions in which subjects had some idea as to how well they were performing tasks.

Dr. Patrick Lawther of St. Bartholomew's, London, and his group also have been studying the effects of central nervous system exposures in their laboratory. Their subjects show little reaction to CO exposures, possibly again because they are engaged competitively in tasks which they are capable of evaluating and overcompensating for as they work.

The session chairman at the New York Academy meeting, Arthur DuBois of the University of Pennsylvania, emphasized the importance of experimental design. He noted that the effects of CO appear to be those which could readily be overcome with attention. Therefore, to test the true diminution of function, one needs first to replicate Beard's experimental design before it is possible to evaluate the effects of different experimental systems on central nervous system function. It seems that even for such a common substance as CO we shall still learn of new environmental effects.

Monitoring Carbon Monoxide in Ambient Air

R. S. Ajemian and N. E. Whitman

Monitoring Carbon Monoxide in Ambient Air

The interest in ambient carbon monoxide levels has been increasing in recent years. This interest has received considerable impetus from the establishment of air quality standards by various governmental agencies and from the purely academic interest in the sources of CO generation and its removal from the atmosphere.[1] Due to this increased interest in ambient CO levels, an investigation into the techniques used for monitoring CO was undertaken.

Although various methods such as iodine pentoxide and indicator tubes have been used for detection purposes, the method commonly employed for the detection and monitoring of CO in ambient air has been limited to the use of long path infrared instruments. Infrared instruments provide a simple and reliable means of analyzing recording data over prolonged periods. These instruments, however, require the use of specialized equipment in order to achieve a sensitivity greater than 0.5 ppm.

Recently a highly sensitive instrument utilizing the reducing power of carbon monoxide on mercuric oxide and the subsequent detection of mercury vapor by a mercury vapor detector was described by Robbins, *et al.*[2]

Instrumentation of this kind is relatively costly. Purchase of more than one such instrument normally required to monitor an area of interest for an alert or background purposes would create a budgetary problem for most organizations. Therefore, the solution to the problem was to utilize equipment which would be as inexpensive as possible so that several locations could be monitored simultaneously.

Monitoring Equipment

In order to overcome the prohibitive cost of purchasing instrumentation described above, an attempt was made to adapt our existing equipment for the monitoring and detection of carbon monoxide. To achieve this end, we utilized a sequential sampler commonly used for monitoring various other pollutants in ambient air over a 24-hr period. The air sampled each hour by the sequential sampler was collected in plastic bags and returned to the laboratory for analysis using a gas chromatograph with a flame ionization detector. Since laboratories devoted to the analysis of atmospheric samples usually have a gas chromatograph and sequential samplers, no additional cost is incurred in such a program for monitoring ambient carbon monoxide levels (Figure 1).

A sequential sampler manufactured by the Research Appliance Co. which can collect twelve consecutive samples by means of a timer programmer and solenoid valves was used. The sampling interval of this programmer can be varied over a period of 12–45 min to obtain 12 air samples over a 24-hr period. After the twelfth sample, the sequencer will turn off the sampler automatically. The sampler was used by setting the timer for 12 min and cutting the airflow back to approximately 0.30 l/min. This was accomplished by placing a needle valve in the line from the pump which was adjusted empirically to obtain the desired airflow. The approximately 3.5 liters of air sampled is more than adequate for analysis. The prime concern in the use of a sampler of this type is its ability to operate at a reduced rate of flow without damaging the pump. Low flow rates are required because of the handling problem which could occur by sampling large volumes of air in the plastic bags.

Plastic Bags

The plastic bags were made in our laboratory by heat sealing two 12 in. × 12 in. pieces of Scotchpak® film with a tire valve in one side of the bag. We have found that Mylar and polyethylene bags are also suitable for this purpose. When Mylar is used, a special thermoplastic tape was used for sealing the edges.* The bags can be used for storage of the air samples for 1 week without any apparent loss in carbon monoxide concentration.

During the course of sampling, about 48 bags are required for the monitoring program. When the samples are collected, 12 bags are required for sampling purposes and 12 are needed for replacement purposes. The extra 24 bags we found were helpful for standby purposes should some bags be lost or should the analyses of the samples be delayed for a day or so.

Analysis of Sample

The bag samples collected by the sampler for carbon monoxide can be analyzed by the silver salt of *p*-sulfaminobenzoic acid[3] or by gas chromatography. We found that by use of 1-liter sample bottle the silver salt technique will provide reliable results; however, it is very time consuming due to the 2-hr shaking period required. Using the 1-liter bottle, approximately 0.5 ppm of carbon monoxide can be detected.

In order to avoid the long period of shaking, a gas-chromatographic method for carbon monoxide was adopted. In the method described by Porter and Volman,[4] carbon monoxide which produces very little response in the hydrogen flame, is converted catalytically to methane by mixing the carrier steam with the hydrogen and passing it over a special nickel catalyst on fire-brick packed into a 5 in. stainless steel tube operated at 285°C. The column used for fractionation was 10 ft of ¼ in. stainless steel pack with 35 × 48 mesh Linde 5A molecular sieve[4] which we had previously prepared for other purposes. Our Perkin-Elmer Model 154 was modified according to Figure 2.

Using a 25-ml loop we are able to detect about 0.1 ppm of carbon monoxide on our chromatograph. Quite frequently during the analysis the flame

---

* Bags may be purchased from Calibrated Instruments, Inc., 17 West 60th St., New York, N. Y. Tire valves may be purchased from Halkey-Roberts Corp., Spring Valley Rd , Paramus, N. J.

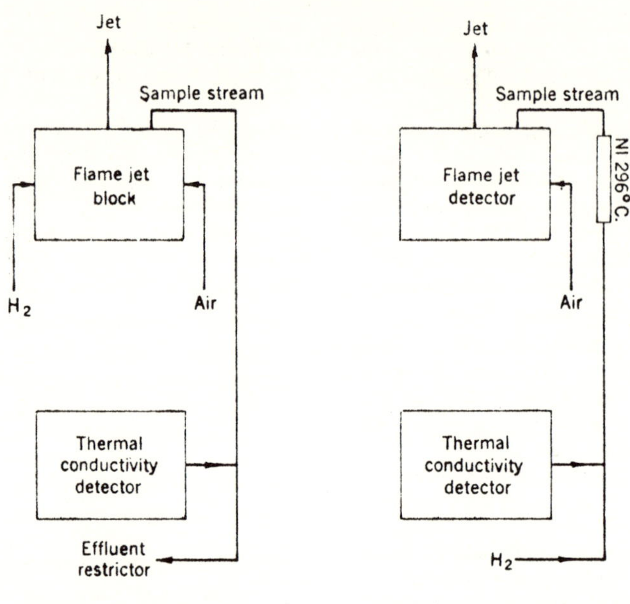

Figure 2. Diagram showing flow modification of Perkin-Elmer Model 54

blows out because of the large volume of air used for the sample. However, the carbon monoxide peak is well separated from the false peak created by relighting the flame. With higher carbon monoxide concentrations, a smaller loop (5 ml) can be used and no problem with the flame is encountered.

Calibration curves were prepared using low levels of carbon monoxide made by the dilution of pure carbon monoxide with air. An allowance was made for the carbon monoxide in air by subtracting the peak height obtained for air only from that made by the curve for carbon monoxide in the 1–10 ppm range. Using pure methane as a comparison, we found that the catalyst converted the carbon monoxide quantitatively to methane. The time required to analyze and calculate 12 samples is approximately $2\frac{1}{2}$ hr.

Summary

A method was devised for the determination of carbon monoxide in ambient air utilizing the equipment normally found in the laboratory. Use was made of a sequential sampler by reducing the airflow and collecting samples over a 15-min period in plastic bags containing approximately 3.5 liters. The samples were analyzed by the catalytic conversion of carbon monoxide to methane and its subsequent detection by the flame ionization detector. Should a gas chromatograph not be available, the silver salt of *p*-sulfaminobenzoic acid can be used; however, the analysis would require considerably more time.

References

1. Jaffe, L. S., "Ambient carbon monoxide and its fate in the atmosphere," *J. Air Poll. Control Assoc.*, **18**, 534 (1968).
2. Robbins, R. C., Borg, K. M., and Robinson, E., "Carbon monoxide in the atmosphere," *J. Air Poll. Control Assoc.*, **18**, 106 (1968).
3. Levaggi, D. A., and Feldstein, M., "The colorimetric determination of low concentrations of carbon monoxide," *Am. Ind. Hyg. Assoc. J.*, **25**, 64 (1964).
4. Porter, K. and Volman, D. H., "Flame ionization detection of carbon monoxide for gas chromatographic analysis," *Anal. Chem.* **34**, 748 (1962).

Motor Vehicle Emissions

Composition, Size and Control of Automotive Exhaust Particulates

G. L. Ter Haar, D. L. Lenane, J. N. Hu, and M. Brandt

The Federal Government recently proposed a 1975 standard for particulate emissions from vehicles. Although we are now faced with a new parameter in vehicle emissions, we are not aware of any definitions concerning automotive particulates nor do we know of any methods set forth to date to measure total particulates.

Several years ago, studies of lead emissions from vehicles under several driving conditions showed that:

1. Only a portion of the lead burned in the fuel is exhausted; and
2. The percentage of lead that is exhausted becomes less as operating conditions become lighter.[1,2]

A review of the available published information on particulate emission gives some idea of automotive contribution to total man-made atmospheric particulate. The contribution of the automobile varies with particulate load from values as low as 3% in a "heavy particulate" city (St. Louis) to 37% in a basically clean area (Los Angeles), with the average for the United States being about 1.8%.[3,4] NASN data for the lead content of total suspended atmospheric particulates shows, for 27 cities where data are available, that only 0.51% of the particulate is lead. Thus, it seems we are being asked to control, at a large cost, a very small portion of the total atmospheric particulates.

Emission and Composition of Exhaust Particulates

Two important questions arise concerning the proposed automotive particulate standard. These are: "What are the particulate emissions of cars now?" and "What portion of these particulates is lead compounds?" A summary of information available in the literature is shown in Table I.

These data are from the Department of Health, Education, and Welfare studies made in the last few years.

"

	Total, g/mi	Lead, % of total
1962 Sumner Tunnel study[5]	0.36	8.6
1966 Total filter study (8 cars)[6]	0.31	...
1967 Morse Report[7]	0.7[a]	19.0
1969 HEW air qual. crit. (12 lb/1000 gal)[8]	0.36[b]	...

[a] Based on assumed total vehicle mileage.
[b] Based on assumed fuel economy of 15 miles/gallon.

The only study in which lead was actually determined was the Sumner Tunnel study in 1962. This work was done by government scientists who measured particulates in the inlet and outlet air of the tunnel and concluded that the increase was due to vehicle traffic. They observed that the suspended particulates in the tunnel were produced at a rate of 0.36 g/mile and that lead represented 8.6% of the total.

For many years, Ethyl Corporation has studied the lead portion of exhausted particulates. As early as 1925, vehicle tests were conducted under road operation using an electrostatic precipitator. Hirschler, *et al*, reported in 1957 and 1964 on the results of an extensive chassis dynamometer study of exhausted lead, including particle-size analysis.[1,2] As a portion of our continuing study of automotive exhaust products, we began to look for methods of measuring the total particulate emission of vehicles.

Test Procedure

Our current procedure is one that does not measure total particulates from a vehicle but is a step in this direction. We are using a large black polyethylene bag to collect exhaust gases and suspended particulates from a vehicle operated on the 7-mode Federal emission test cycle. The particulates are sampled from the bag by filtration methods for analysis. Early studies with bag sampling techniques indicated that an air dilution factor of at least 8:1 was necessary in order to minimize the effect of humidity on particle agglomeration rate and fallout. This meant we needed a bag with a volume in excess of 2500 cu ft, since approximately 300 cu ft of exhaust would be developed by the engine during four Federal 7-mode cycles. Black polyethylene is used in the manufacture of the bag in order to avoid any photochemical reactions that might aid in the formation of secondary particulate. The vehicle is operated on a Model C-250 Clayton chassis dynamometer with exhaust ducted into the bag during Federal 7-mode cycle operation. Both air and fuel consumption are measured during test running.

Since lead can be detected much more easily than other particulate matter, several initial tests of lead distribution were made to determine the degree of mixing that had occurred by natural air currents. We found that suspended lead concentration was essentially the same at all locations in the bag and picked two sampling points at the 5-foot height level and in the horizontal center of one of the 28-foot long walls.

Prior to admitting exhaust into the chamber, the bag is flushed with room air. This diluent air is then circulated by an internal blower system through a dehumidifier. After the moisture content reaches a minimum, the air is again circulated through a large bed of Drierite followed by a MSA-HEPA absolute filter. This treated diluent air has undetectable lead content and a relative humidity of less than 10%. The vehicle is then operated on the chassis dynamometer from either a cold start or with a fully warmed-up engine. The exhaust from four Federal 7-mode cycles is introduced into the bag.

As soon as the vehicle operation is complete, two sampling systems are activated. One uses a $\frac{1}{4}$ in. stainless steel probe that extends approximately 6 ft into the bag. This probe remains in the bag and is connected to a 47-mm Millipore holder mounted on the exterior of the bag wall. Lead concentration samples are obtained on Type AA 0.84 pore Millipore filters at a flow rate of 0.25 cfm. The second sampling system employs an open-face Millipore XX50-047-10 holder that has been modified to eliminate flow restrictions. This holder is inserted into the bag through the other sample port in the bag, and the sample flow rate is adjusted to 2 cfm. A 4.25-cm glass filter paper is used in this holder for determination of total particulate and carbon concentrations. Both sampling systems employ positive displacement meters for determination of sample volume.

Lead content on the Type AA Millipore filter is determined either by wet chemical methods or with an atomic absorption spectrometer. Total particulate is determined on the glass filter by weight difference, after the filter is equilibrated at 50% RH, using a Mettler microbalance. Carbon is determined by burning the glass filter sample in a Leco furnace. Analytical data are reported in micrograms per cubic foot of gas sampled. Engine air and fuel consumption data are used to determine total exhaust volume, which enables us to calculate a dilution ratio for bag mixture. The total air-suspended particulate matter, lead, and carbon can then be calculated, with data expressed in terms of grams per mile.

Our early work was done measuring only lead concentration and the decay of this concentration with time. For example, we found that, for a typical hot-cycle test, lead concentration was halved in about 18 to 24 hr. Initially, about 18% of the lead burned was exhausted. After 24 hr, only 7% of the lead burned remained suspended. It is possible to calculate particle size distribution when a concentration-time curve is available.[9,10] We believe this is a rather crude method of particle sizing, but it did indicate a stable aerosol of unit density would decay similarly if the particles were aerodynamically about 1 micron in size. The effect of aerosol concentration also was investigated by introducing one to four cycles in the bag.

The settling curves had nearly identical slopes for a dilution range of 10:1 to 40:1, indicating that, at least in this concentration range, wall losses and fallout were not affected by exhaust gas concentration.

Our incentive for building this system was to develop a technique for measuring and characterizing the types of particulates being produced by vehicles under cyclic driving conditions. In addition, we wanted to be able to evaluate the effectiveness of our vehicle particulate traps on exhausted particulates other than lead. We recover lead chemically from vehicle engines and exhaust systems and relate the total recovery to the lead consumed by the engine during the test period. Although this method gives us evaluations of trapping systems in terms of lead, we find very little material in the deposits that is not associated with lead compounds. For example, any carbon or oil that might be trapped would probably decompose in the exhaust system during high-temperature operation. Even in cars with high oil consumption, we have not found a large amount of carbon or lube oil residues in the trap, while other more stable compounds are trapped and ultimately recovered. For example, a component we found in the trap systems during winter operation was sodium, which occurred in concentrations in excess of 10% by weight and was probably a result of street salting.

We felt that a sampling procedure was needed for measuring total particulates and carbon. To obtain sufficient samples for weighing, we first attempted to use the ¼ in. sampling line system at a flow rate of 2 cfm. Data were very erratic, and lead concentration appeared to be very low. A flow study indicated that the entry section of the Millipore holder was an excellent trap at the high flow rate. As previously discussed, we constructed a sampling system that could use the 47-mm Millipore paper, but would have an unrestricted entry. Early tests were run with both sys-

Table II. Suspended-particulate emission and composition—stabilized deposits (30,000–100,000 mi).

Federal cycles	No. of Deter-minations	Particulates, g/mi			% Pb	% C
		Total	Pb	C		
4 Cold	16	0.512	0.085	0.184	16.4	35.9
4 Hot	17	0.240	0.044	0.076	18.3	31.7
Weighted[a]	16	0.339	0.059	0.115	17.4	33.9

[a] Weighted 35% cold and 65% hot.

tems, the 1/4 in. probe and the enlarged holder, sampling simultaneously. Lead analysis indicated that the total-filter probe was measuring a representative sample of the exhaust-air mixture.

Results

We have conducted a large number of tests using the black-bag technique. Early experiments involved studies designed to determine concentration profiles, settling rates, and reproducibility. Recently, we have concentrated on measuring total suspended particulates and the analysis of these particulates for lead and carbon. Several trapping systems also have been checked during this period of time.

We did not have baseline cars paired with our trap vehicles to give us a comparison of trapping effectiveness on the small suspended particulates. Therefore, emissions were obtained from 16 representative cars that had been driven in normal service using leaded commercial gasolines. These cars were mainly 1966 pre-emission-control vehicles representing 5 makes and 7 engine types. The average deposit mileage for the 16 cars was 50,444. Table II summarizes these results.

The weighted total suspended emissions for the 16 determinations, both hot and cold cycles, averaged 0.339 g/mi. The emissions during the 4 warm-up (cold) cycles were more than double the hot-cycle emissions. This ratio varies with condition of the vehicle and probably is affected by choke operation. One of the cars tested had cold-cycle emissions of 1.2 g/mi, while the warmed-up (hot) emissions were 0.161 g/mi, or a ratio of 7.5 to 1. This points out the significant effects that driving and vehicle conditions can have on measured emission rates. Lead emission for these cars averaged 0.059 g/mi for the composite weighted cycle, or 17.4% of the total particulates.

The lead data also can be expressed in terms of lead compound by calculating emission rates as the $PbClBr$ form. Analysis of the lead particulate indicates that most of the exhusted lead appears as $PbClBr$. On this basis, the emission rate for lead compounds for the weighted cycle is 0.092 g/mi or 27.1% of the total particulates. Carbon represented about 34% of the total (weighted) emissions, being 0.115 g/mi —somewhat greater than the emission of lead compounds. Lead emission also can be related to the amount of lead consumed by the engine. For the group of cars tested, the lead emitted was 26.3% of the lead burned. This level is in about the same range as reported by Hirschler in 1964[2] and estimated from data reported by Mueller in 1962.[11]

A few cars were available that had been operated in owner-type service solely on unleaded commercial premium gasoline for at least 30,000 mi. We have investigated particulate emissions of five such cars using the bag technique. Data for these stabilized-deposit, unleaded-fuel cars are shown in Table III.

These data show that these stabilized

Federal cycles	No. Deter.	Particulates, g/mi		
		Total	C	% C
4 Cold	5	0.316	0.242	76.5
4 Hot	5	0.134	0.074	55.2
Weighted[a]	5	0.197	0.133	67.5

[a] Weighted 35% cold and 65% hot.

unleaded-fuel cars emitted about 40% less total particulates than the leaded-fuel cars (Table II). Carbon represented nearly 70% of the total. A comparison of cold-cycle carbon emission rates with and without lead indicates that leaded fuel reduces the carbon emission during cold-start and warm-up conditions.

We also have been determining suspended-particulate emissions from many new test cars after an initial break-in using unleaded Indolene. The cars also are tested using Indolene 30 as they are assigned to various test programs in our laboratories. Table IV summarizes the results of these tests.

Federal hot-cycle data are available for a large number of cars, but cold-cycle data have been obtained on only 6 leaded-fuel cars and 3 unleaded-fuel cars. In new cars, there is little difference in suspended total particulates. This reflects the high trapping efficiency of a clean exhaust system. An average of 7.3% of the lead burned is exhausted by these low-mileage cars.

Probe Sampling

The bag technique appeared to be relatively complex, and the number of tests that could be made was limited by bag clean-up and dehumidification. We felt that a quick method of particulate measurement would be a direct probe of the tailpipe at a constant flow rate, which could then be related to actual particulate found in the large bag. We used a 47-mm filter holder and infrared lamps to keep the holder above the condensation temperature of the exhaust. A non-proportional constant-flow sample was drawn from the tailpipe at a flow rate of 1 cfm. Total particulate was determined by weight on a microbalance using the same procedure as was used with the bag samples.

The results of these tests are compared with bag results in Table V. These data show that particulate from unleaded-fuel cars is difficult to sample even with absolute Millipore-type filters. Certainly, a probe sampling system underestimates by a large factor the actual amount of particulate that is exhausted.

We have made several tests using an absolute filter on a car. The normal emissions from one representative car operated on four Federal 7-mode warm-up cycles were 0.344 g/mi total particulates, 0.070 g/mi lead, and 0.133 g/mi carbon. A large 24 in. × 24 in. × 12 in. MSA filter was installed on the tailpipe of the vehicle, and the test was repeated. Bag concentrations were reduced to 0.010 g/mi total particulates, 0.0 g/mi lead, and 0.003 g/mi carbon. This test showed that over 95% of the total particulates, all of the lead, and over 95% of the carbon were filterable. We conclude that very little particulate is forming in the bag during the time required to fill and complete the sampling tests. Therefore, we believe that serious difficulties are inherent in a sampling technique that obtains particulate matter from a hot-flowing system.

Fuel Effects

Part of our research program involved study of particulate emissions with several fuel additives in a common base fuel. These tests were made under hot-cycle conditions in a 1970-model car (Make D) equipped with a 350-CID V-8 engine after accumulation of 5000 deposit miles. Figure 1 shows the relative effect of several additives compared to baseline operation on 3.0 ml TEL/gal as Motor Mix. Unleaded Indolene showed a 37% reduction in particulate matter. The addition of a carburetor detergent to this fuel caused

Table IV. Suspended-particulate emissions from new cars (0–1000 mi).

Federal cycles	No. of Determinations	Particulates, g/mi			% Pb	% C
		Total	Pb	C		
		Leaded-Fuel Cars				
4 Cold	6	0.204	0.026	0.111	12.7	54.4
4 Hot	15	0.117	0.019	0.040	16.2	34.2
Weighted[a]	6	0.152	0.020	0.071	13.2	46.7
		Unleaded-Fuel Cars				
4 Cold	3	0.223				
4 Hot	22	0.107				
Weighted[a]	3	0.165				

[a] Weighted 35% cold and 65% hot.

a small increase in total suspended particulates, while phosphorus and a commercial upper cylinder lubricant caused substantial increases in particulate emission.

Fuel composition also plays an important role in particulate emissions. We have studied the effect of the aromatic content of unleaded gasoline on suspended carbon emission from a 1970 Make D car equipped with a 350-CID V-8 engine. Two fuels were blended to give a range of aromatic content from 10 to 40%, about the range of current commercial gasolines. The aromatic-free fuel was a full-boiling alkylate, and the high-aromatic fuel was a commercial unleaded premium. Figure 2 shows that suspended carbon emission using the composite Federal cycle doubled as aromatic content increased from 10 to 40%.

Conclusions

Studies of suspended particulate emissions from several cars operated on a chassis dynamometer under Federal 7-mode cycle conditions have shown that:

1. Cars vary widely in their particulate emissions.
2. Cold-cycle operation gives 2 to 8 times more particulates than hot-cycle operation.
3. Lead compounds represent less than one-third of the total particulates.
4. Carbon emission for deposit-stabilized cars varies widely, but averages about 35% of the total for leaded-fuel cars and nearly 70% for unleaded-fuel cars.
5. Suspended particulate emissions are nearly equal with new cars whether or not lead is present in the fuel.
6. Exhausted lead varies with the condition of the exhaust system, but ranges between 7 and 30% of the lead consumed by the engine.
7. Fuel composition and additives affect the amount of emitted particulates.
8. Probe-sampling techniques underestimate by a large factor the amount of particulates emitted by cars.

Lead Particulates

There has been a continuing interest in the size distribution of lead particulate exhausted from the automobile. Hirschler, *et al.*,[1] determined the lead exhausted under cycling driving conditions. They found that about 50% of the lead burned was exhausted during city driving. Higher speeds, accelerations, and decelerations increased the amount of lead emitted. During city driving, less than 5% of the exhausted lead had an actual geometric size of less than 1μ. Heavy particles 5μ and larger made up 27% of the exhausted lead, with the remainder being between 1 and 5μ. Hirschler used a precipitator to collect the total particulates from the exhaust during cruising and cycling conditions. The particulates were then redispersed in a liquid, and their sizes were determined. This method of sizing may have influenced the size of the particulates.

Mueller, *et al.*,[11] found that, when a car was operated under cruise conditions 62–80% of the lead particulate exhausted was less than 2μ in size. More than 68% of these small particles were less than 0.3μ. Mueller's work was done at constant speed, which is not representative of actual driving conditions. Under the driving conditions used in his study, about 10–$20\mu g$ of lead per liter were exhausted. Assuming that about 100 μg of lead per liter of exhaust would be emitted if all the lead burned were exhausted, only 10–20% of the lead burned was exhausted in Mueller's study.

Habibi[12] has reported on an elaborate system using a huge mixing tunnel. This method allows for proportional sampling of the exhaust from a cycling driving schedule. It also allows for direct determination of aerodynamic size in impactors. He studied a car operated on an AMA cycle and a 45-mph steady state. Under the steady-state conditions, he obtained a mass median diameter (MMED) of about 0.6μ. Under the conditions of the AMA cycle, the MMED increased with mileage. The MMED was about 1μ at 7250 miles, about 2μ at 16,500 miles, and about 4μ at 21,150 miles.

Although this latter method avoids many of the problems associated with the first two methods, it is very expensive to set up and much time and effort are needed to obtain a single result.

Our goal was to sample many automobiles under various driving conditions in a way that would not influence the size of the exhausted particulate. Initially, we used an Andersen Sampler (a six-stage impactor) with a probe inserted into the tailpipe to sample directly. A Millipore filter was used to collect all particles that went through the Andersen Sampler.

Results of direct-probe sampling from the tailpipe with an Andersen Sampler during 30-mph cruise and Federal Cycle conditions showed that the exhaust lead particles were either collected on the first stage or the seventh stage (final filter). When the sample was collected behind a cyclone, nearly all of the lead was on the final filter. This means that there was very little lead particulate in the size range represented by Stages 2 through 6 (5μ to 0.5μ in aerodynamic equivalent diameter of unit density). The exhaust lead particles were either greater than 5μ or less than 0.5μ in diameter under these conditions.

The sub-micron fraction ($<0.5\mu$) probably comprises primary particles freshly produced during the combustion process. The coarse fraction ($>5.0\mu$) consists of secondary particles that have deposited on the walls of the exhaust system and are exhausted due to mechanical and thermal shocks. These particles are large enough to settle out quite quickly from the atmosphere.

Because of these observations, we decided to use a sampling system that would divide the particulate into a fine and a coarse fraction and would allow us to sample under varying gas flow.

Experimental

A 3-in. Aerotec cyclone followed by an 8 in. × 10 in. filter (shown in Figure 3) was used to sample the total exhaust gas and to separate the exhaust particles into two fractions. The cyclone-filter combination was connected directly to the tailpipe of a car. Since the collection efficiency of the cyclone varies with flow rate [D_{50} (size of the particles at which the cyclone in 50% efficient) equals 5μ at 30 ft³/min and 0.6μ at 60 ft³ min],[13] we did not know exactly the 50% cut-off point (D_{50}) of the cyclone in a Federal Cycle. Fortunately, our Andersen Sampler results showed that only a small amount of the exhaust lead particulate was in the 0.5 to 5μ range. Thus, using a cyclone did not introduce much error in the results.

We used the Clayton dynamometer and the chassis dynamometer for the tests. The cyclone and the filter holder were preheated with infrared lamps to prevent water vapor from condensing. The exhaust gas was directed into the

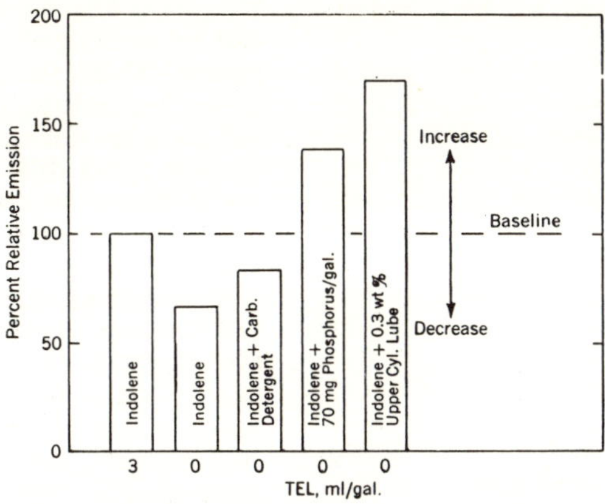

Figure 1. Effect of fuel additives on particulate emissions. 1970 make D, 4 FTP hot cycles.

sampling line for a period of time. After a sample had been taken, the exhaust gas was directed into the bypass, the cyclone collection can (Mason jar) and the filter were removed, and the lead content was determined.

The sampling time varied with driving conditions. In most cases, the sampling time was 5 min for 30-mph and 45-mph cruise and 2 min for 60-mph cruise. The seven-mode Federal Cycle runs were single-cycle tests. Under these conditions, the exhaust gas flows were relatively low and 8 in. $\times$ 10 in. filters were used for the tests. During the 0–70 mph or the 30–70 mph wide-open-throttle (WOT) acceleration tests, the 8 in. $\times$ 10 in. filter was too small to handle the extremely high exhaust flow. Therefore, a large MSA (10-in. diameter) pleated filter was used for these tests.

For cold-start tests in the Federal Cycle, the cars were left standing overnight and the amount of lead exhausted during the first four cycles was collected, using the cyclone-filter combination. For the hot-cycle portion of the Federal Cycle, the cars were warmed up for 25–30 min before sampling.

Results

We have tested 26 cars—10 cars owned by the Ethyl Corporation in commuter service and 16 employee-owned cars. The Ethyl cars were 1963–68 models that had been driven from 20,000 to 62,000 miles. The employee cars were all 1966 models with 17,000 to 92,000 miles of service.

The test conditions included cruise at 25, 30, 45, and 60 mph, the 7-mode Federal Cycle, and WOT accelerations of 0–70 and 30–70 mph. The results, expressed as percent lead burned, were averaged for each car for each driving condition.

The 26-car average results are summarized in Table VI. These data show that, as the conditions under which the car is operated become more vigorous, the percentage of lead burned that is exhausted increases and the coarse particles become a larger percentage of the exhausted lead. At 25-mph cruise, only 12% of the lead is exhausted, and over 90% of this is fine particles. At 60-mph cruise, 33% of the lead is exhausted, with one-third being coarse and two-thirds being fine.

The results for the hot-cycle portion

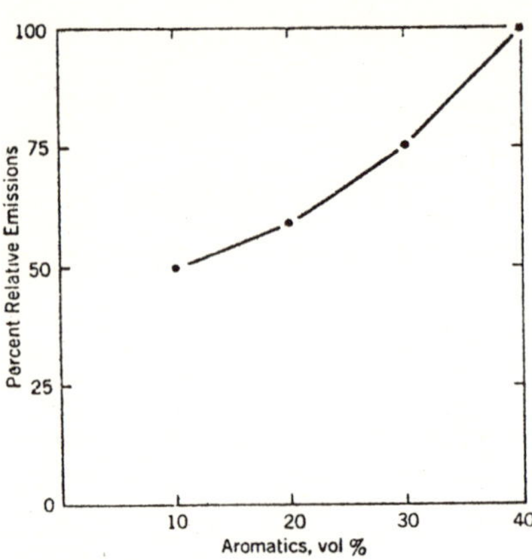

Figure 2. Effect of fuel aromaticity on suspended carbon particulate emission. 1970 make D, composite Federal cycle.

of the Federal Cycle tests lie in between these extremes: 21% of the lead burned is exhausted, with 29% being coarse and 71% being fine. These data for the hot-cycle portion of the Federal Cycle only account for about one-fifth of the lead burned. Hirschler, *et al.*,[1] found that about 75% of the lead burned was exhausted over the lifetime of the car.

To account for the remainder of the lead, cars were run under more-vigorous conditions, including the cold-cycle portion of the Federal Cycle and WOT accelerations. The cold-cycle data show about 44% of the lead burned was exhausted, with approximately equal amounts of fine and coarse particles. The highest percentages of lead were emitted during WOT accelerations. When accelerations were made up to 70 mph, about 1100% of the lead was emitted, with two-thirds coarse and one-third fine.

Using these data and assuming as shown by Hirschler and co-workers that 25% of the lead remains in the automobile, we have made some calculations on the distribution of coarse and fine particles. First, about 21.4% of the lead burned was emitted on the average from 26 cars on the Federal hot cycles—

15.2% was fine and 6.2% coarse. The cold cycles emitted 43.9% of the lead burned—24.4% fine and 19.5% coarse. Since the Federal Cycle is weighted 65% for the hot cycles and 35% for the cold cycles, the calculated percentage of burned lead emitted during the full Federal Cycle is 29.2%–18.4% fine and 10.8% coarse.

If we add the 29% for the full Federal Cycle to the 25% we estimate remains in the car, about 54% of the lead burned is accounted for. We then have assumed that the remaining 46% is emitted during heavy accelerations and decelerations. The data in Table VI show that, under these conditions, 35% of the lead particles emitted are small and 65% are large. Using these data, size distribution of the total lead exhausted can be calculated. For the fine particles, $18\% + (35\% \times 0.46) = 34\%$. For the coarse particles, $11\% + (65\% \times 0.46) = 41\%$. Thus, 55% of the exhausted lead is coarse particles.

As 55% of the lead is larger than 5μ, the MMED is also larger than 5μ, which is in agreement with results obtained by Habibi.[12] During cruise conditions, the MMED is smaller than 0.5μ, which is consistent with the work of Mueller and co-workers.[11]

We conducted a 7-hr test at 25-mph cruise. A 3 in. cyclone and a 10 in.-diameter MSA pleated filter were used to collect the exhaust lead. Of the lead burned, 10.9% was collected in the filter ($<0.5\mu$) and only about 1% was collected in the cyclone. The results confirmed those of our 5-minute tests, as well as Mueller's results.

We also have run a long Federal Cycle test. The same combination of cyclone and MSA filter was used. The car was continuously driven for 6 hours using the Federal Cycle. The results showed that about 20% of the lead burned was exhausted, with about equal amounts of coarse and fine particles.

Total Filter Car

To confirm the work in the laboratory, we equipped a car with a 3 in.

Table V. Comparison of sampling systems—federal hot cycles.

Car			Total particulates, g/mi		
Year	Make	Test Fuel	Probe	Bag	Ratio
Leaded-Fuel Cars					
1968	R	Ind. 30	0.025	0.149	5.9
1970	B		0.036	0.124	3.4
1970	B		0.054	0.189	3.5
Unleaded-Fuel Cars					
1968	B	Comm.	0.0036	0.087	24.3
1968	B		0.0029	0.069	24.0
1966	A	Prem.	0.0110	0.230	20.9
1966	D		0.0024	0.120	50.0

Aerotec cyclone and a 24 in. × 24 in. × 12 in. MSA total filter on the tailpipe. The cyclone-filter combination was mounted in the trunk.

Road tests showed that the pressure drop across the filter was only 0.3–0.4 in. H_2O during WOT acceleration from 0 to 60 mph and 0.1 in. H_2O during 60-mph cruise.

Unleaded fuel was used for these road tests. Just before the lead-balance test was started, the fuel was changed to Indolene + 3 ml TEL. The car was driven on a modified AMA route to accumulate mileage.

The 50% cut-off point for the cyclone was about 0.5 to 5μ, depending on the flow through it. The absolute filter had a 99.9% efficiency for 0.3μ particles.

The particles caught in the cyclone were considered to be coarse, while those that passed through and were caught in the filter were considered to be fine. The cyclone can was changed every 3000 mi, and the filter was changed at 6000 and 12,000 mi (end of test). Oil and the oil filter also were changed at 6000 and 12,000 mi. At the end of the test, the engine cylinders were scraped and the deposits were analyzed for lead. The exhaust system was cut up, scraped, and soaked in an EDTA solution. The total filter also was soaked in EDTA to extract the lead.

Table VII shows the results of this experiment. We accounted for 84.3%

of the lead burned. We believe that the 15% unaccounted for was lost during handling of the exhaust system and the total filter. These are both extremely difficult systems to handle and to obtain total recovery of lead.

The coarse fraction of the lead burned (cyclone) was 31.8% overall, but it was 49% during the last 3000-mi increment. The fine fraction (filter) was 22.5% of the lead overall and was higher in the second 6000 mi (27.5%). This is expected in both cases as the 20% of the lead found in the exhaust system builds up during the earlier mileage. By 9000 mi, it was probably near equilibrium. The oil and oil filter caught a constant 10% overall.

Although the total lead exhausted overall was about 55% of the lead burned, it is instructive to note that it was only 36% for the first 6000 mi and increased to 73% for the last 6000 mi. As long as the exhaust system is not changed, it seems likely that this car will continue to exhaust 75% or more of the lead burned.

The results of this test were very similar to our observations in the laboratory. These results also support Habibi's observation that the MMED of lead particulate emitted became larger as mileage was accumulated on the car.[12]

Conclusions

We concluded from the tests using the full Federal Cycle that 18% of the lead burned is emitted as fine particles and 11% as coarse; these represent emission rates of 0.051 and 0.028 g/mi respectively. Over the lifetime of the cars, we estimate that about 35% of the lead burned is emitted as fine particles and 40% as coarse. We found similar results on a car on which a total lead balance was made. This car also demonstrated that exhausted lead particles become larger as mileage accumulates.

Other investigators have shown that most of the lead particles emitted during steady-state conditions are fine, but the quantity of lead emitted is only a few percent. Our work agrees with this observation. Earlier work also has shown that the average size of the lead particles and the percentage of lead exhausted increase with the severity of the operating conditions. This work also confirms these observations.

The method described in this paper is quick, relatively inexpensive, and allows sampling under cycling conditions. It is especially well suited to a survey of the lead emissions from a large number of cars. The method checks well with the work of other investigators when the method of operation of the car is taken into consideration.

Particulate Traps

Earlier, it was stated that we were interested in a technique that would help us evaluate our trapping systems in terms of effectiveness on suspended particulate emissions. Our current procedures evaluate trapping effectiveness on only lead compounds, since we re-

Table VI. Lead exhausted during various driving conditions.

| | Exhausted Lead | | | | | |
| | % of Pb burned | | | g/mi | | |
	Coarse	Fine	Total	Coarse	Fine	Total
Driving conditions						
25-mph cruise	0.9	10.8	11.7	0.002	0.027	0.029
45-mph cruise	3.9	14.4	18.3	0.010	0.036	0.046
60-mph cruise	10.3	22.4	32.7	0.026	0.057	0.083
Federal cycle						
Hot cycles	6.2	15.2	21.4	0.014	0.042	0.056
Cold cycles	19.5	24.4	43.9	0.054	0.067	0.121
Full[a]	10.8	18.4	29.2	0.028	0.051	0.079
WOT Acceleration[b]	733	386	1119			

[a] Weighted 35% cold and 65% hot.
[b] 0–70 and 30–70 mph.

Table VII. Total lead balance study.

| | Percent of Pb burned | | | | |
Miles	Oil and filter	Cyclone	Total filter	Exhaust system	Total
3000		13.6			
6000	10	18.2	17.7		
		(23.2)[a]			
9000		25.7			
		(42.0)[a]			
12,000	10	31.8	22.5	20	84.3
		(49.0)[a]	(27.5)[a]		

[a] Incremental.

126

cover collected material from the units at end-of-test and then chemically recover lead. Performance is based on the lead retained as a percentage of total lead burned. Since we have "lead balanced" several standard exhaust-equipped vehicles, we are able to relate trap performance in terms of reduction of lead emission to the environment. Reductions of 65% are attained with simple inertial designs that can directly replace a standard muffler. More complex trapping systems can reduce exhausted lead by 80–95%.[14]

An example of a relatively simple inertial device is shown in Figure 4. This unit is composed of two inertial elements that we call anchored vortex tubes. Particulate separated at the walls is rejected at high energy through the slots near the base of the closed-end tube. One of these units was operated on a 1969 Make D car and reduced exhausted lead by 65% for 30,000 mi. The effects of this prototype trap on emission of suspended total particulates and lead are shown in Table VIII for three car makes after 6000 mi of normal driver operation. More complex traps based on interception of small particles, agglomeration of these particles, and release from the interceptor section for removal in an inertial system have shown reductions in suspended total particulates in excess of 70%. Overall reductions in exhausted lead with these systems are greater than 90%. These results indicate that a simple inertial trap has the ability to reduce effectively exhausted lead in the suspended size range and that more sophisticated traps will substantially reduce both suspended lead and total particulate matter.

References

1. Hirschler, D. A., Gilbert, L. F., Lamb, F. W., and Niebylski, L. M., "Particulate lead compounds in automobile exhaust gas," *Ind. Eng. Chem.*, **49**: 1131 (July 1957).
2. Hirschler, D. A. and Gilbert, L. F., "Nature of lead in automobile exhaust gas," *Arch. Environ. Health*, **8**: 297 (Feb. 1964).
3. "Air Quality Criteria for Particulate Matter," U. S. Dept. of Health, Education, and Welfare, *NAPCA Publication No. AP-49*, Jan. 1969.
4. "Nationwide Inventory of Air Pollutant Emissions—1968," U. S. Dept. of Health, Education, and Welfare, *NAPCA Publication No. AP-73*, August 1970.
5. Larsen, R. I. and Konopinski, V. J., "Sumner Tunnel air quality," *Arch. Environ. Health*, **5**: 597 (Dec. 1962).
6. Hangebrauck, R. P. Lauch, R. P., and Meeker, J. E., "Emissions of polynuclear hydrocarbons from automobiles and trucks," *Amer. Ind. Hyg. Assoc. J.*, **27**: 47 (Jan.-Feb. 1966).
7. "The Automobile and Air Pollution: A Program for Progress," Report of the Panel on Electrically Powered Vehicles, U. S. Dept. of Commerce, Oct. 1967.
8. "Control Techniques for Particulate Air Pollutants," U. S. Dept. of Health, Education and Welfare, *NAPCA Publication No. AP-51*, Jan. 1969.
9. Dimmick, R. I., Hatch, M. T., and NG, J., "A particle sizing method for aerosols and fine powders," *AMA Arch. Ind. Health*, **18**: 23 (July 1958).
10. Rose, H. E. and Langmaid, R. N., "Calculation of particle size distribution from sedimentation observations," *Nature*, **199**: 774 (April 1957).
11. Mueller, P. K., Helwig, H. L., Alcour, A. E., Gong, W. K., and Jones, E. E., "Concentration of Fine Particles and Lead in Car Exhaust," *ASTM Special Tech. Publication No. 352*, 60–73, December 1963.
12. Habibi, K., "Characterization of particulate lead in vehicle exhaust—Experimental techniques," *Environ. Sci. Technol.*, **4**: 239 (1970).
13. Freudenthal, P., "High collection efficiency of the Aerotec-3 cyclone for submicron particles," to be published.
14. Hirschler, D. A. and Marsee, F. J., "Meeting Future Automobile Emission Standards," *NPRA Paper No. AM-70-5*, April 1970.

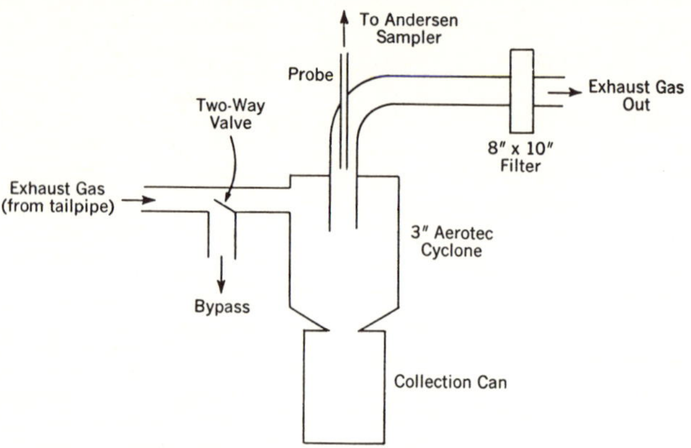

Figure 3. Schematic of sampling system.

Figure 4. Experimental dual anchored vortex trap.

Table VIII. Particulate trapping—Federal 7-mode cycle.

Car Year and Make	Miles	Exhaust systems	Particulates, g/mile Total	Pb
1970 make D	6000	Standard	0.169	0.046
	6000	Anchored vortex	0.139	0.039
		% Reduction	19	16
1970 make B	6000	Standard	0.274	0.071
	6000	Anchored vortex	0.227	0.056
		% Reduction	17	21
1970 make A	6000	Standard	0.471	0.114
	6000	Anchored vortex	0.189	0.043
		% Reduction	60	66
1970 make D	0	Standard	0.116	
	0	Agglom.-inertial #1	0.037	0.006
		% Reduction	68	
16 cars (1966–1969)	30,000+	Standard	0.339	0.060
1965 make D	24,000	Agglom.-inertial #2	0.125	0.036
		% Reduction	63	39
1969 make D	12,000	Agglom.-inertial #3	0.108	0.034
		% Reduction	67	43
1970 make D	12,000	Agglom.-inertial #4	0.098	0.019
		% Reduction	71	68

CONCENTRATION AND PARTICLE SIZE DISTRIBUTION OF PARTICULATE EMISSIONS IN AUTOMOBILE EXHAUST

Robert E. Lee, Jr., Ronald K. Patterson, Walter L. Crider and Jack Wagman

Abstract—Particulate components in hot, concentrated automobile exhaust and in cooled, diluted exhaust were fractionated and collected with cascade impactors; analyses were made for sulfate, nitrate, chloride, and water-soluble and water-insoluble lead. All components analyzed were contained in particles predominantly submicron in diameter. About 95 per cent of total lead was associated with particles having aerodynamic diameters below 0.5 μm. The particle size distributions of sulfate and nitrate were dependent on engine operating conditions, the particle size decreasing with increasing road speed and exhaust temperature. Photoirradiation produced an increase in the concentrations of water-soluble lead, sulfate, nitrate, and chloride particulate but a decrease in the average particle size of sulfate, nitrate, and chloride. After photoirradiation nearly all chloride was in the particle size range below 1 μm dia. Two mechanisms, both based on photochemical transformations of lead compounds, are suggested as explanations for a relatively lower accumulation of lead in the bones of mice exposed to irradiated, as compared to nonirradiated, exhaust.

INTRODUCTION

MANY of the gaseous constituents that form the visibility-obscuring suspended particulates in photochemical smog are emitted by motor vehicles; such substances include carbon monoxide, reactive hydrocarbons, and oxides of nitrogen and sulfur (LARSEN, 1966). Furthermore, many of the particulate components identified in auto exhaust emissions are present in ambient air in a predominantly respirable range: the mass median diameters of particulate lead, sulfate, and nitrate in urban air range, with a few exceptions, from 0.2 to 0.6 μm (ROESLER *et al.*, 1965; WAGMAN *et al.*, 1967; LUDWIG and ROBINSON, 1968; LEE and PATTERSON, 1969; ROBINSON and LUDWIG, 1967; LEE *et al.*, 1968). What fraction of these components is contributed by the motor vehicle is uncertain. Since the particle size distribution of particulate emissions can affect the degree of pulmonary penetration and retention (DAUTREBANDE, 1962), visibility reduction in the atmosphere (MIDDLETON, 1952), and a host of particle–particle and particle–gas interactions, it is important to define the concentration and size distribution of these components in motor vehicle exhaust.

Although studies of the concentration and size distribution of particulates in automobile exhausts were reported by HIRSCHLER (1957), McKEE and McMAHON (1960), and MUELLER *et al.* (1963), the available data are still meager. In the atmosphere, auto exhaust components undergo extensive changes through photochemical and other transformations. For example, lead halide aerosols can undergo photolysis (PIERRARD 1969) and aerosols can be formed photochemically from gaseous mixtures of hydrocarbons, nitrogen oxide, and sulfur dioxide (STEVENSON *et al.*, 1965). LUTMER *et al.* (1967) found that concentrations of lead in the bones of mice exposed to nonirradiated diluted auto-exhaust were higher than in mice exposed to comparable dilutions of irradiated auto-exhaust. This finding suggests that irradiation produces lead particulate products that are absorbed differently than are nonirradiated lead components.

In the work reported here, the concentrations and particle size distributions of lead, sulfate, nitrate, and chloride were measured in hot, undiluted auto-exhaust and in cooled auto-exhaust diluted with clean air. Some effects of photoirradiation on the particulate and gaseous components of diluted auto-exhaust, and the influence of added sulfur dioxide, were also determined.

EXPERIMENTAL METHODS

Particulate emissions were fractionated in six stages of inertial separation and collected on stainless steel plates with an Andersen* cascade impactor (ANDERSEN, 1965) operating at 1 ft^3 min^{-1}. A type AA Millipore membrane filter, 105 mm in diameter, was placed immediately behind the last stage to collect unimpacted particles. The material collected on each stage and on the backup filter was chemically analyzed. Cumulative particle size distribution curves were obtained, as in previous studies (WAGMAN et al., 1967, LEE et al. 1968), by log-normal plots of stage D_{50} values (determined for the Andersen sampler by FLESCH et al. 1967), against mass cumulative per cent for each component.

Hot undiluted auto-exhaust

An Andersen cascade impactor with backup filter was mounted in a 20-gal galvanized steel chamber, shown in FIG. 1. Raw undiluted auto-exhaust from a 6-cylinder 1965 Plymouth burning regular-grade Gulf gasoline was passed directly from the tail-pipe into the sampling chamber. The automobile was operated on a chassis dynamometer at idle, 30 m.p.h. and 50 m.p.h. and the exhaust was sampled for 15 min after the system reached a stable temperature. The sampler was stoppered until chamber temperature reached equilibrium and unstoppered from the outside (FIG. 1) immediately before sampling.

The Andersen sampler, heated by the exhaust gases, collected particulate emissions without condensing an appreciable amount of water. Tests showed that the sampler could be operated at temperatures as high as 150°C without significantly altering the distance from jet to collection surface or destroying the membrane backup filter or the rubber gaskets mounted between the stages. After a sample was collected, the chamber was cooled slowly to room temperature before the sampler was removed. Condensed water was found on the bottom of the chamber but not on the sampler or other internal fixtures.

Diluted auto-exhaust

Auto exhaust that was cooled and diluted about 1:600 with clean air was sampled with Andersen cascade impactors from a series of animal inhalation exposure chambers for 16 h. Although the auto-exhaust source, dilution and photoirradiation procedures, chamber construction, and preparation and monitoring of exposure atmospheres have been described by HINNERS et al. (1966, 1968), they are outlined briefly in the following paragraph.

* Mention of commercial products does not constitute endorsement by the Public Health Service.

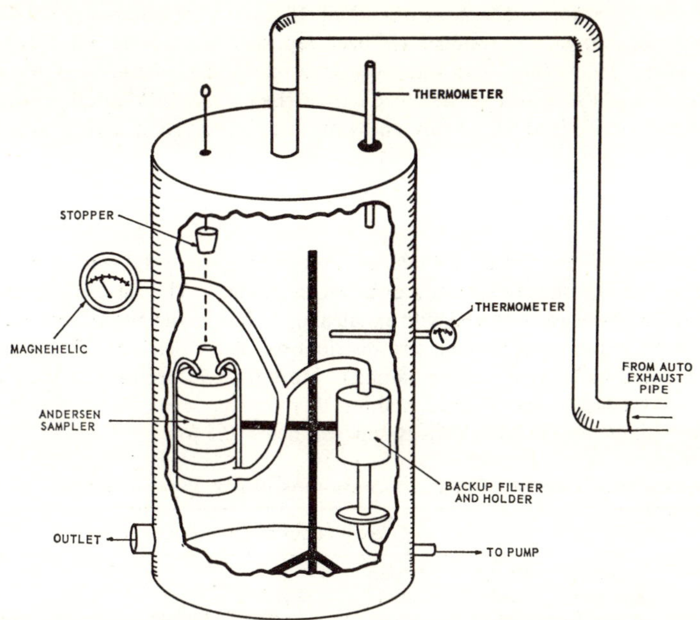

Fig. 1. Apparatus for sampling hot, undiluted auto-exhaust directly from an automobile exhaust pipe.

Exhaust from an automobile engine coupled to a dynamometer unit, and operated in a 5-min cycle similar to the California cycle (HINNERS, 1963), was proportionally diluted with clean conditioned air under regulated pressure and fed to dynamic flow photoirradiation chambers. The photoirradiation chambers (about 680 ft^3 vol. each) were illuminated by a composite of three types of fluorescent tubes designed to simulate sunlight. These covered the wavelength range 3000–6000 Å with peaks at about 3100 Å, 3500 Å, and 4500 Å; average residence time was about 1 h (HINNERS, 1963). The dilution system also provided nonirradiated exhaust in the same concentration. The exhaust, after cooling and irradiation, was passed at a flow of 11 ft^3 min^{-1} to a series of animal exposure chambers from which samples were taken for analysis. Mixtures of auto-exhaust and sulfur dioxide, added after the auto-exhaust passed through the irradiation chamber, were also used. Each chamber was monitored for carbon monoxide, hydrocarbons, nitrogen dioxide, nitric oxide, and ozone by methods described by HINNERS *et al.* (1966). Some detailed hydrocarbon analyses were made by gas chromatography.

Chemical analysis

The stainless steel Andersen plates and backup filters were initially extracted with hot water as described earlier (WAGMAN *et al.*, 1967). The water extracts were analyzed for sulfate (WAGMAN *et al.*, 1967); for nitrate (SALTZMAN, 1954) following reduction to nitrite with hydrazine sulfate; and for chloride (BERGMANN and SANIK, 1957). It was later learned that the method used for chloride involves an interference by bromide (Zall *et al.*, 1956). Since the stoichiometry of the interference is not known, the results have been reported as chloride but should be regarded as halide mixture. The water-insoluble particulate matter was re-extracted with 10 per cent nitric acid as described earlier (Lee *et al.*, 1968); both water-soluble and water-insoluble lead were analyzed by atomic absorption spectrophotometry.

RESULTS

Lead

The concentrations of water-soluble, water-insoluble, and total lead in raw exhaust sampled at various driving speeds are shown in Table 1. The ratio of water-soluble to water-insoluble lead increased with exhaust temperature and speed from 2.42 at idle to 6.45 at 50 m.p.h. The particle size distribution of the lead could not be determined in detail with the Andersen sampler. More than 95 per cent of the measured lead passed the last impactor stage, and was therefore associated with particles having equivalent diameters (for unit density spheres) below 0.5 μm.

Table 1. Concentrations of particulate components in hot, undiluted raw auto-exhaust

Auto speed	Ave. sampling chamber temp. (°C)	Concentrations of components, μg m^{-3}					
		Water-sol. lead	Water-insol. lead	Total lead	Sulfate	Nitrate	Chloride
idle	48.5	7580	3103	10,683	774	252	2876
30 m.p.h.	65.0	2884	1081	3956	884	696	2179
50 m.p.h.	125.0	5969	915	6883	291	409	3641

The average lead concentrations for three consecutive samples in cool, diluted auto-exhaust are given in Table 2. In the nonirradiated chambers, more than 70 per cent of the total lead was water-soluble. Irradiation increased the proportion of water-soluble lead to 85 per cent, with a corresponding decrease in the amount of water-insoluble lead. A similar effect was observed with the addition of sulfur dioxide. Again, detailed size distributions for lead particulate in the chambers could not be determined with the Andersen sampler, since about 95 per cent was associated with particles below 0.5 μm dia.

Sulfate

The sulfate concentration in hot, concentrated auto-exhaust diminished appreciably during engine operation at 50 m.p.h., at which speed the exhaust temperature in the sampling chamber averaged 125°C (Table 1). The size distribution curves for particulate sulfate (Fig. 2A) also exhibited an apparent dependence on sampling temperature, wherein particle sizes decreased with increasing engine speed and exhaust temperature.

Sulfate concentrations were low in nonirradiated, diluted exhaust; however, irradiation produced a 16-fold increase. In chambers with diluted exhaust and added sulfur dioxide, irradiation produced a 4-fold increase in sulfate concentration, accompanied by a decrease in the average sulfur dioxide concentration from 40.1 to 34.2 pphm.

TABLE 2. AVERAGE CONCENTRATIONS OF GASEOUS AND PARTICULATE COMPONENTS IN COOLED, DILUTED AUTO-EXHAUST

	Nonirr. exhaust	Irr. exhaust	Nonirr. exhaust + SO_2	Irr. exhaust + SO_2
Particulate, μg m^{-3}				
Water-sol. Pb	8.8	10.2	10.7	11.5
Water-insol. Pb	3.7	1.7	2.8	2.4
Total lead	12.5	11.9	13.5	13.9
Sulfate	0.37	5.9	21.0	99.2
Nitrate	3.3	581	11.3	223
Chloride	20.2	28.7	17.6	32.7
Gases, ppm				
SO_2			0.40	0.34
O_3	trace	0.25	trace*	0.35
NO	1.96	0.20	1.96*	0.09
NO_2	0.04	0.88	0.04*	0.77
CO	100	94.4	100*	98.3
HC	35.5	23.6	35.5*	24.9

* Estimated from nonirradiated exhaust chambers.

Average size distribution curves for particulate sulfate in the diluted exhaust (FIGS. 2B and 2C) show a pronounced shift to smaller particle sizes with irradiation, suggestive of the formation of small sulfate particles from sulfur dioxide. Irradiation increased the proportion of total particulate sulfate associated with submicron particles from about 75 per cent to 90 per cent.

Nitrate

The particulate nitrate concentration in hot, raw exhaust varied with temperature and operating speed (TABLE 1) in a manner similar to that observed for sulfate. The size distribution of nitrate (FIG. 3A) also showed a shift to smaller particle sizes with increasing exhaust temperature and operating speed. At 50 m.p.h., 99 per cent of the nitrate passed the last Andersen stage, and was therefore contained in particles of less than 0.5 μm dia.

Large quantities of particulate nitrate were formed by irradiation of diluted exhaust (TABLE 2). This effect was preceded by a sharp decrease in the NO concentration and a substantial increase in the NO_2 levels. The addition of 0.5 ppm sulfur dioxide to diluted exhaust increased appreciably both the concentration (TABLE 2) of nitrate and the sizes (FIGS. 3B and 3C) of nitrate-containing particles. This effect is unexpected and cannot be explained at this time. As observed also for sulfate, irradiation produced a shift of nitrate to smaller particle sizes with and without added sulfur dioxide. On the other hand, the addition of sulfur dioxide to diluted irradiated exhaust decreased the amount of NO_2 and nitrate formed by irradiation, probably due to competitive inhibition from sulfate formation.

Chloride

The concentration and particle size distribution of particulate chloride in hot, concentrated auto-exhaust varied with exhaust temperature and operating speed in the manner shown in TABLE 1 and FIG. 4A. From 79 to 94 per cent of the chloride was associated with particles that were less than 1 μm dia.

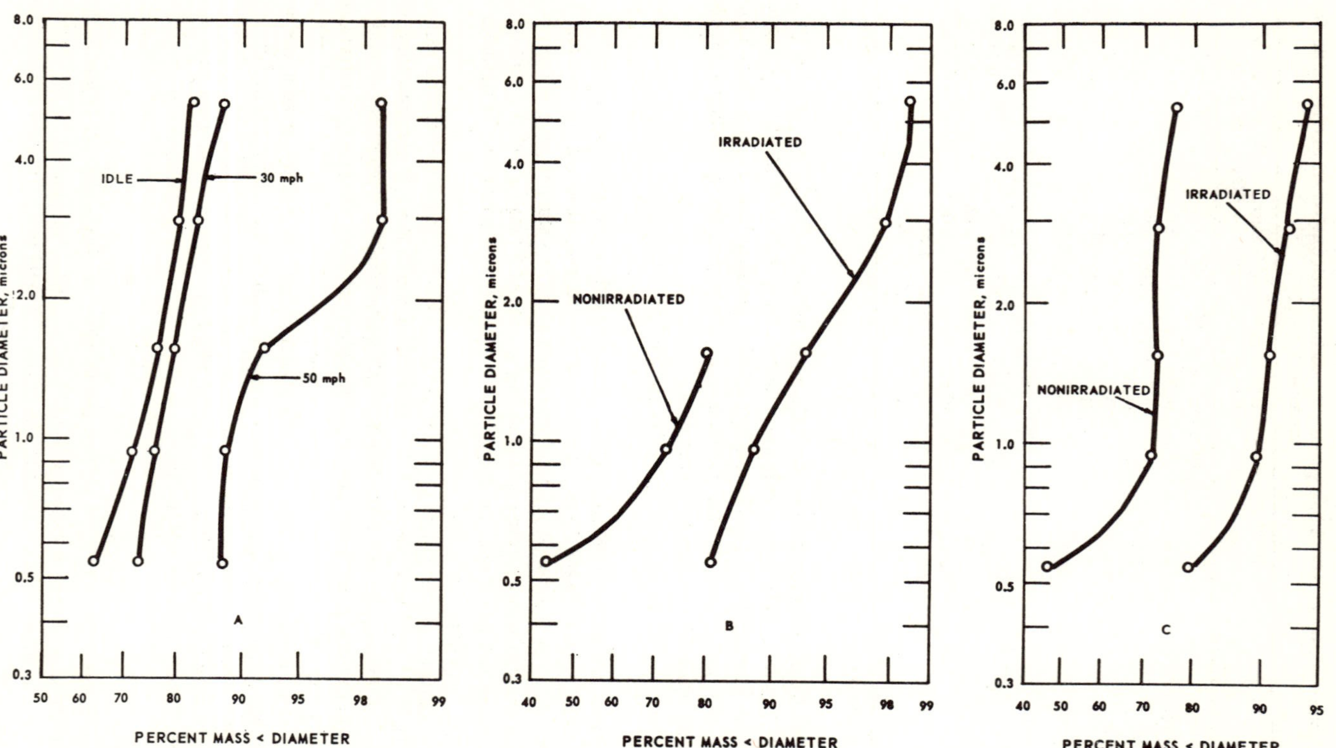

Fig. 2. Cumulative particle size distributions of sulfate in (A) hot, undiluted auto-exhaust at various operating speeds; (B) cooled auto-exhaust diluted 1:600 with clean air; (C) cooled auto-exhaust diluted 1:600 with clean air and mixed with 0.5 ppm sulfur dioxide.

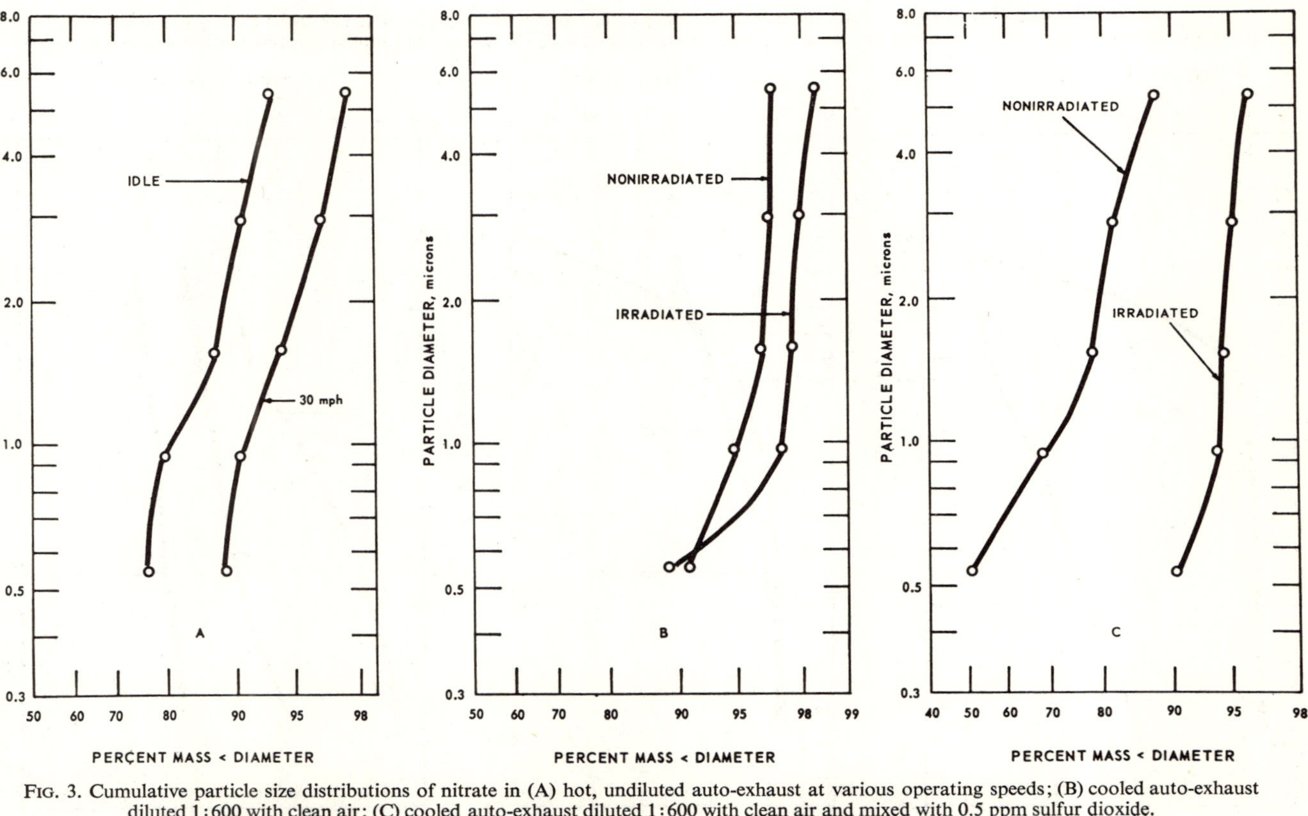

FIG. 3. Cumulative particle size distributions of nitrate in (A) hot, undiluted auto-exhaust at various operating speeds; (B) cooled auto-exhaust diluted 1:600 with clean air; (C) cooled auto-exhaust diluted 1:600 with clean air and mixed with 0.5 ppm sulfur dioxide.

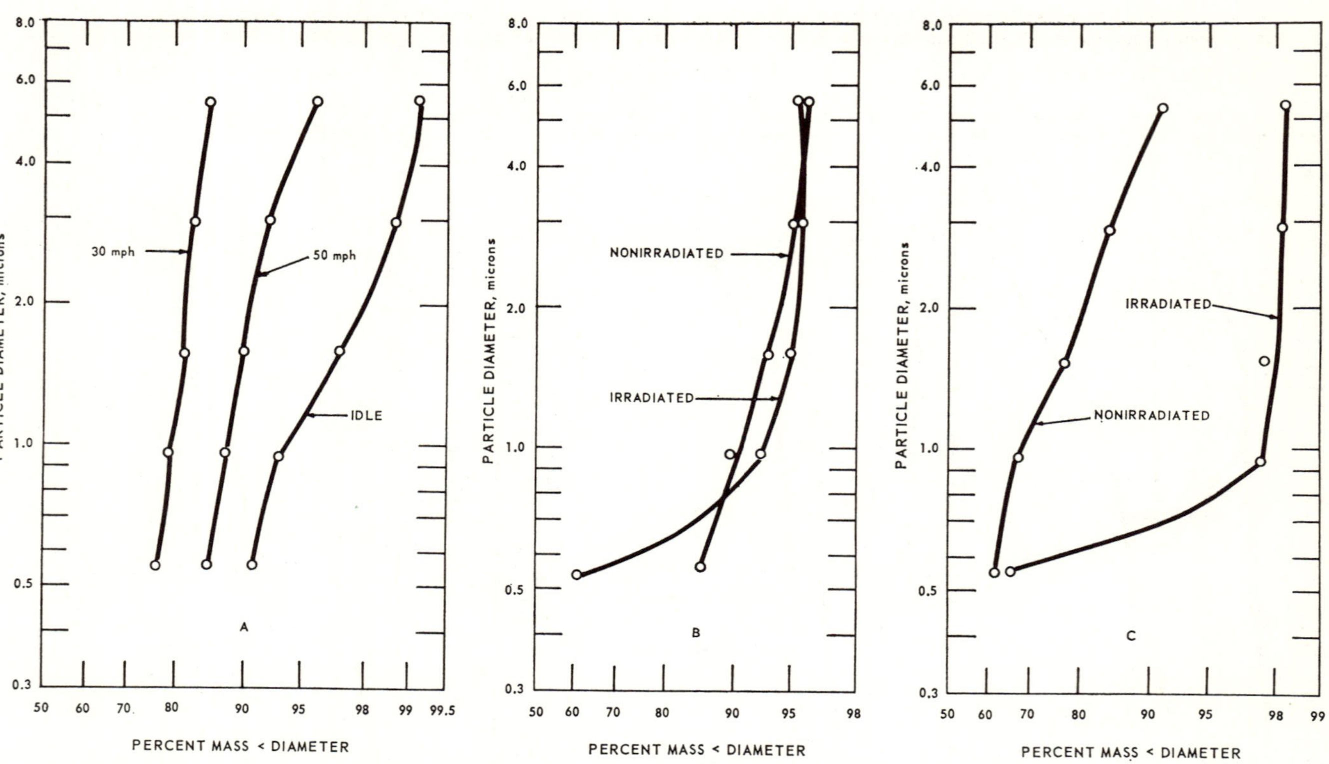

Fig. 4. Cumulative particle size distributions of chloride in (A) hot, undiluted auto-exhaust at various operating speeds; (B) cooled auto-exhaust diluted 1:600 with clean air; (C) cooled auto-exhaust diluted 1:600 with clean air and mixed with 0.5 ppm sulfur dioxide.

Irradiation of diluted auto-exhaust increased the particulate chloride concentration (TABLE 2) and produced changes in both the range and shape of the chloride size distribution curves (FIGS. 4B and 4C). The effect on particle size was especially pronounced in the presence of added sulfur dioxide, when irradiation increased the proportion of total chloride in particles below 1 μm dia. from about 70 per cent to 98 per cent without an appreciable change in the proportion of chloride in particles below 0.5 μm dia.

DISCUSSION

The concentration of water-insoluble lead in hot, concentrated automobile exhaust decreased with increasing exhaust temperature (TABLE 1); this decrease suggests that some of the lead compounds are vaporized at the higher temperatures and are therefore not collectible as particles. None of the other components analyzed showed this monotonic relationship between concentration and exhaust temperature over the entire range of operating conditions.

The high ratio of water-soluble to water-insoluble lead in raw exhaust is not unreasonable, since organic halide compounds are usually added as scavengers to gasoline; much of the exhausted lead is therefore in the form of water-soluble lead halides (HIRSCHLER et al., 1957). The water-insoluble lead probably includes elemental lead, lead oxide, and organic lead. The relatively high concentration and submicron size of particulate lead in automobile exhaust are consistent with atmospheric measurements (ROBINSON and LUDWIG, 1967; LEE et al., 1968). The finding in this study that more than 95 per cent of the lead passed the last (i.e. sixth) impactor stage of the Andersen sampler, and was therefore associated with particles less than 0.5 μm dia., is comparable to the data of MUELLER et al. (1963), who reported about 80 per cent by weight of particulate lead having similar aerodynamic behavior.

In contrast to these controlled-environment studies, measurements under equilibrium conditions in ambient air indicate that the proportion of water-soluble particulate lead amounts to only about 10–15 per cent of the total lead sampled (ROBINSON et al., 1963; ROBINSON and LUDWIG, 1964).

The observation that the proportion of water-soluble lead in exhaust increases with irradiation at the expense of water-insoluble lead is interesting in view of the effect of lead emission on concentrations of lead in the bones of mice. LUTMER et al. (1967) reported that less lead was found in the bones of mice exposed for 15 months to low levels or irradiated auto-exhaust than in corresponding groups exposed to equal concentrations of non-irradiated auto-exhaust. The question arises as to whether the lower lead content is associated with the transformation of specific lead compounds induced by irradiation. Irradiation produces increased concentrations of sulfate and nitrate (TABLE 2), presumably from the oxidation of SO_2, NO and NO_2 followed by reaction with water vapor.

The acid droplets formed could conceivably react with elemental and organic lead to yield quantities of lead sulfate and nitrate, which may account for at least some of the increase in water-soluble lead. LUTMER's observations of increased lead accumulation in the bones of mice exposed to nonirradiated exhaust may be explained by (1) probable differences in the hygroscopicity of various lead aerosol compounds and hence their ability to penetrate the lower respiratory region (MILBURN et al., 1957), and/or (2) an alteration in the lead compounds through photochemical reactions, resulting in substances having less affinity for bone tissue.

The size distribution curves for particulate nitrate and sulfate in raw exhaust were clearly dependent on road speed and exhaust temperature. FIGURE 5 shows how the per cent of total nitrate and sulfate in particles with diameters of 1 μm or less increases with exhaust temperature.

The formation of large amounts of nitrate and sulfate with irradiation is accompanied by a shift to smaller particle sizes for these components. The newly formed particles are smaller than those present before irradiation since they have had little chance to grow by coagulation or condensation of water vapor. The predominantly submicron particle size of these components in automobile exhaust is consistent with data obtained in sampling urban atmospheres (WAGMAN *et al.*, 1967; LUDWIG and ROBINSON, 1968) indicating mass median diameters in the range of 0.2–0.6 μm.

It seems reasonable that aerosol components other than sulfuric and nitric acids are formed by irradiation of SO_2, NO, NO_2 in a complex hydrocarbon mixture. STEVENSON *et al.* (1965) reported that considerable quantities of aerosol can be formed by irradiation of SO_2 with *trans*-2-butene and by irradiation of NO_2 with cyclohexene.

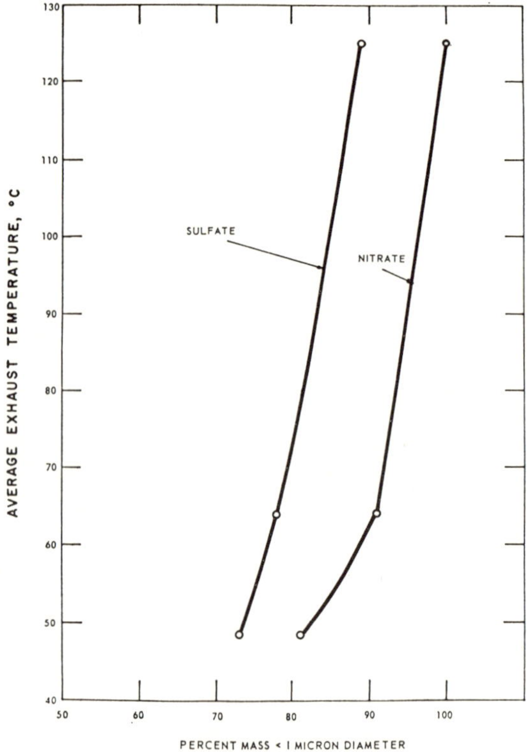

FIG. 5. Variation of the per cent of total sulfate and nitrate in particles smaller than 1 μm dia. with average exhaust temperature.

Such reactions were likely in the experiments reported here, especially in view of the observed reduction of unsaturated hydrocarbon gases with irradiation, shown in TABLE 3.

Engine operating mode had no apparent effect on the particle size distribution of chloride in raw exhaust. In nonirradiated diluted exhaust, addition of sulfur dioxide resulted in a pronounced increase in the sizes of chloride-containing particles; the fraction of total chloride in particles 1 μm dia. or less decreased from 91 to 69 per cent. It is interesting to observe the radiation-induced change in the chloride particle size

TABLE 3. AVERAGE HYDROGEN CONCENTRATIONS (PPM) IN IRRADIATED AND NONIRRADIATED AUTO EXHAUST

	Nonirradiated	Irradiated
C_2H_4–C_2H_6	0.64	0.57
C_2H_2	0.38	0.38
C_3H_8	0.04	0.04
C_3H_6	0.12	0.07
i–C_4H_{10}	0.05	0.05
n–C_4H_{10}	0.16	0.15
l–butene	0.04	Trace
iso-butylene	0.05	Trace
1,3-butadiene	0.08	Trace
i–C_5H_{12}	0.16	0.14
n–C_5H_{12}	0.11	0.11

distributions, especially in the presence of added sulfur dioxide. This was characterized by a marked shift from a broad distribution in particles above 1 μm dia. to a relatively narrow distribution in particles below this size. Apparently the chloride-containing particles formed photochemically are of relatively uniform size in contrast to the broad range of sizes observed for photochemically-formed nitrate and sulfate. This suggests that the photochemically-produced chloride undergoes a process of particle growth that is quite different from that of nitrate and sulfate.

SUMMARY

Particulate samples of hot, concentrated automobile exhaust were fractionated and collected with cascade impactors in a specially designed sampling chamber; samples of cool, diluted auto-exhaust were also taken, and some were exposed to photo-irradiation. Lead, nitrate, sulfate, and chloride components in exhaust emissions were predominantly of submicron size, a finding that is consistent with atmospheric measurements. In hot, concentrated automobile exhaust more water-soluble than water-insoluble lead was found. The particle-size distributions of sulfate and nitrate showed a monotinic temperature-dependence, the average sizes decreasing with increasing exhaust temperature.

With irradiation the concentration of water-soluble lead increased and the concentration of water-insoluble lead decreased. Two possible mechanisms, both based on photochemical transformation of lead compounds, are proposed as explanations for the relatively lower amount of lead that accumulates in the bones of mice exposed to irradiated exhaust.

A.E. 5/4—D

Irradiation produced an increase in the concentrations of particulate chloride, sulfate, and nitrate but a decrease in the average particle sizes. The particle-size distribution of chloride in diluted exhaust was influenced by added SO_2, while irradiation produced a narrow distribution of chloride in particles below 1 μm dia. with and without added SO_2.

Acknowledgements—The authors wish to thank Mr. JEROME FLESCH for his assistance in some phases of the sampling, and Mr. GARY LEWIN, Mrs. BETTIE FAHRLENDER, and Mr. ROBERT SLATER for the chemical analyses.

REFERENCES

ANDERSEN A. A. (1965) A sampler for respiratory health hazard assessment. *Am. Ind. Hyg. Ass. J.* **27**, 160–165.

BERGMANN J. G. and SANIK J. (1957) Determination of trace amounts of chlorine in naptha. *Analyt. Chem.* **29**, 241–245.

DAUTREBANDE L. (1962) *Microaerosols.* Academic Press, New York.

FLESCH J. P., NORRIS C. H. and NUGENT A. E. (1967) Calibrating particulate air samplers with monodispersed aerosols: application to the Andersen cascade impactor. *Am. ind. Hyg. Ass. J.* **28**, 507–516.

HINNERS R. G. (1963) Laboratory-produced automobile exhaust facility. *Biomed. Sci. Instrum.* **1**, 53–58.

HINNERS R. G., BURKART J. K. and CONTNER G. L. (1966) Animal exposure chambers in air pollution studies. *Archs. envir. Hlth* **13**, 609–615.

HINNERS R. G., BURKART J. K. and PUNTE C. L. (1968) Animal inhalation exposure chambers. *Archs. envir. Hlth* **16**, 194–206.

HIRSCHLER D. A., GILBERT L. F., LAMB F. W. and NIEBYLSKI L. M. (1957) Particulate lead compounds in automobile exhaust gas. *Ind. engng Chem.* **49**, 1131–1142.

LARSEN R. I. (1966) Air pollution from motor vehicles. *Ann. N.Y. Acad. Sci.* **136**, 277–301.

LEE R. E. and PATTERSON R. K. (1969) Size determination of atmospheric phosphate, nitrate, chloride and ammonium particulate in several urban areas. *Atmospheric Environment* **3**, 249–255.

LEE R. E., PATTERSON R. K. and WAGMAN J. (1968) Particle size distribution of metal components in urban air. *Environ. Sci. Technol.* **2**, 288–290.

LUDWIG F. L. and ROBINSON E. (1968) Variations in the size distribution of sulfur-containing compounds in urban aerosols, *Atmospheric Environment* **2**, 13–17.

LUTMER R. F., BUSCH K. A. and MILLER R. G. (1967) Lead from auto exhaust: effect on mouse bone lead concentration. *Atmospheric Environment* **1**, 585–589.

McKEE H. C. and McMAHON W. A. (1960) Automobile exhaust particulates—source and variation. *J. Air Pollut. Control Ass.* **10**, 457–461.

MIDDLETON W. E. K. (1952) *Vision Through the Atmosphere.* University of Toronto Press.

MILBURN R. H., CRIDER W. L. and MORTON S. D. (1957) The retention of hygroscopic dusts in the human lungs. *Am. Med. Ass. Archs ind. Hlth* **15**, 59–62.

MUELLER P. K., HELWIG H. L., ALCOCER A. E., GONG W. K. and JONES E. E. (1963) Concentration of fine particles and lead in car exhaust. Symposium on Air Pollution Measurement Methods, *ASTM Special Tech. Publ.* No. 352, 60–77. American Society for Testing and Materials, Philadelphia.

PIERRARD J. M. (1969) Photochemical decomposition of lead halides from automobile exhaust. *Environ. Sci. Technol.* **3**, 48–51.

ROBINSON E. and LUDWIG F. L. (1967) Particle size distribution of urban lead aerosols. *J. Air Pollut. Control Ass.* **17**, 664–668. ROBINSON E. and LUDWIG F. L. (1964) Size distributions of atmospheric lead aerosols. Final Report, SRI Project No. PA-4788, Stanford Research Institute, Menlo Park, California, April 30.

ROBINSON E., LUDWIG F. L., DeVRIES J. E. and HOPKINS T. E. (1963) Variations of atmospheric lead concentrations and type with particle size. Final Report, SRI Project No. PA-4211, Stanford Research Institute, Menlo Park, California, November 1.

ROESLER J. F., STEVENSON H. J. R. and NADER J. S. (1965) Size distribution of sulfate aerosols in the ambient air. *J. Air Pollut. Control Ass.* **15**, 576–579.

SALTZMAN B. E. (1954) Colorimetric microdetermination of nitrogen dioxide in the atmosphere. *Anal. Chem.* **26**, 1949–1955.

STEVENSON H. J. R., SANDERSON D. E. and ALTSHULLER A. P. (1965) Formation of photochemical aerosols. *Int. J. Air Wat. Pollut.* **9**, 367–375.

Wagman J., Lee R. E. and Axt C. J. (1967) Influence of some atmospheric variables on the concentration and particle size distribution of sulfate in urban air. *Atmospheric Environment* **1**, 479–489.

Zall D. M., Fisher D. and Garner M. Q. (1956) Photometric determination of chloride in water. *Anal. Chem.* **28**, 1665–1668.

POLYCYCLIC AROMATIC HYDROCARBONS FROM GASOLINE-ENGINE AND LIQUEFIED PETROLEUM GAS ENGINE EXHAUSTS

V. Del Vecchio, P. Valori, C. Melchiorri and A. Grella

ABSTRACT

Exhaust gases from an internal combustion engine, alternatively powered by gasoline or by liquefied petroleum gas, were analysed for polynuclear aromatic hydrocarbons. The engine operated a lift truck for the handling of warehouse goods. A combined condensation-filtration apparatus was constructed for sampling of the exhausts. The basic steps of the analytical procedure are summarized here. Several polycyclic aromatic hydrocarbons were identified and determined both in the gasoline engine and in the liquefied petroleum gas engine exhausts. In the liquefied petroleum gas engine exhausts, all detected polynuclear hydrocarbons were present in significantly lower concentrations.

INTRODUCTION

Accumulated evidence has indicated that polycyclic aromatic hydrocarbons (PAH) are present in automotive-engine exhausts. This source contributes significantly to urban atmospheric pollution in terms of known or potential carcinogenic factors such as some of the polynuclear compounds[4, 5, 11, 13–15, 18, 19, 21, 23–25, 29, 33]. It has been established that the production of PAH by an internal combustion engine depends upon several variables such as type of fuel used, mileage of engine, number of revolutions per minute (rpm), change of speed, and load under which the engine is running[7, 9, 11, 15, 20, 21]. Several investigations have shown that many of the recovered polynuclear compounds are common to both gasoline and diesel exhausts. However, notable quantitative differences were found[3, 11, 12, 15, 16, 19, 26, 27].

Nonetheless, no references pertaining to similar studies on internal combustion engines powered by liquefied petroleum gas have been found in the literature. This fuel is very little used in autovehicle engines; therefore, the question of its contributing appreciably to air pollution does not exist. However, in several industrial activities, liquefied petroleum gas is often used to power internal combustion engines which operate lift trucks in loading, unloading, moving and stowing goods. Due to the theoretical possibility of attaining conditions which are closer to the ideal air/fuel ratio and, thus, reducing exhaust emissions (especially as regards carbon monoxide and unburned hydrocarbons), the replacement of gasoline by liquefied petroleum gas would be a preventive measure for the workers involved in the operations mentioned above.

This report is concerned with the qualitative and quantitative determinations of PAH in the exhaust gases from a lift truck alternately powered by gasoline and liquefied petroleum gas. The results are compared and discussed. The first section of this paper consists of a description of the sampling apparatus and analytical procedure used.

EXPERIMENTAL

A lift truck with the following characteristics was employed: model Towmotor, 461 HF type, continental engine, 4 cylinders, 2200 cc displacement, peak 2400 rpm, 2000 kg mx load. At the time of these experiments, the lift truck had operated for 8450 working hours. The engine was alternately powered by gasoline (84 NO) and a propane-butane mixture.

The trials on the lift burning liquefied petroleum gas were accomplished under the following conditions:

(a) on a simulated working schedule with load of 1000 kg,

(b) while the engine was idling (500 rpm) at zero load.

The gasoline-powered engine was only subjected to condition (a).

1. Collection of samples

A combined condensation-filtration apparatus was constructed. The system is illustrated in *Figure 1* and a photograph of the apparatus is shown in *Figure 2*.

A glass probe (8 mm in diameter) with a dual series of holes along the sides, and fitted with a container for condensed water, was introduced about 30 cm into the exhaust pipe to withdraw the sampled gases. These were first passed through a condenser, at the bottom of which was attached a second container for condensed water, and then through a series of three traps cooled in 'dry ice' placed in a Dewar flask.

The relatively dry gas was then filtered through a series of three filters: a fibre glass filter (Gelman, type A-DOP 99·95 per cent efficient, 4·5 in. in diameter), followed by two cellulose triacetate filters (Gelman, GA-6, 0·45 microns pore size, 2 in. in diameter). A vacuum pump sampled at a constant rate of 1 CFM. A vacuum gauge was included to correct the indicated flow for error due to pressure drop across the filters. A running time meter indicated cumulative operating time. Teflon connecting tubes were used. The sections of the system were connected by means of glass standard joints. The apparatus guarantees the quantitative recovery of organic materials of high molecular weight and the accurate measurement of the actual gas volume sampled.

2. Extraction of the PAH from condensed water and filters.

After sampling, the entire condensation section was washed with a definite amount of methanol which had been collected, together with condensed water, in the container placed at the bottom of the condenser. The water-methanol mixture was adjusted to obtain the 4:1 ratio between methanol and water, according to Hoffmann and Wynder[10]. This solution was shaken three times with cyclohexane to enrich the PAH in the cyclohexane phase after separation from the hydrophilic compounds. The filters were extracted in a Soxhlet apparatus with cyclohexane for 20–30 hours.

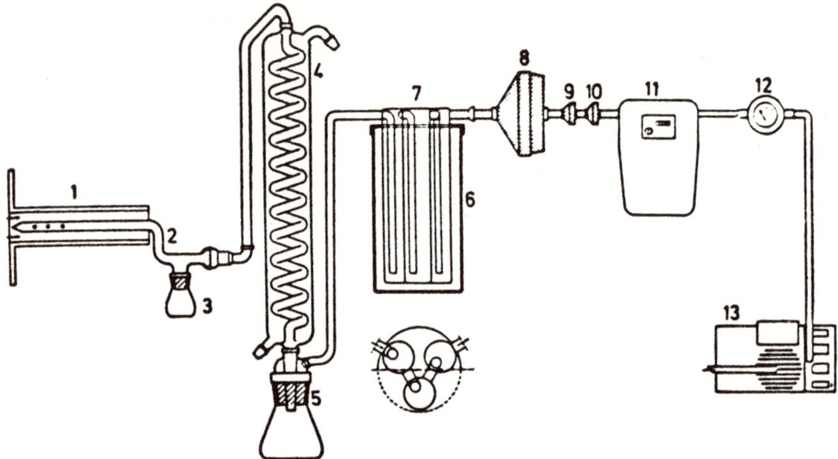

Figure 1. Schematic drawing of the engine exhaust sampling apparatus for determination of polycylic hydrocarbons. (1) Exhaust pipe (2) Probe (3) Condensed water flask (I) (4) Condenser (5) Condensed water flask (II) (6) Dewar flask (7) Traps in dry ice (8) Glass fibre filter (9, 10) Cellulose triacetate filters (11) Dry gas meter (12) Vacuum gauge (13) Vacuum pump.

Figure 2. Engine exhaust sampling apparatus for determination of polycyclic hydrocarbons.

3. Analysis of samples for PAH.

The analytical procedure has previously been described[30-32]. A preliminary adsorption of PAH into a series of superimposed alumina columns, avoids evaporation to dryness and the resulting losses due to the appreciable volatility of polynuclear compounds. This also enables the extracts to be purified from many extraneous materials which give rise to an intense background in the ultraviolet absorption spectra and allows a first separation to be achieved of PAH into two large fractions. The basic steps of the method are summarized below and shown schematically in *Figure 3*.

(*a*) *Adsorption of PAH*. The combined cyclohexane extracts were concentrated to 150 ml in a rotary evaporator, at a reduced pressure and at a temperature of 40°. The residual solution was percolated through a series of three superimposed columns which contained respectively:

(i) aluminium oxide (neutral, Woelm) with 6 per cent of water (Column A);

(ii) aluminium oxide with 3 per cent of water (Column B);

(iii) highly activated aluminium oxide (Column C).

Columns with an outer grinding at the top and an inner grinding at the bottom were used.

The compounds having four or more condensed aromatic rings were retained by alumina placed in column B. Those having a lower number of condensed rings remained adsorbed on the activated alumina of column C. Most of the coloured matter, oxygenated products and other substances with greater chromatographic adsorbability were firmly held on alumina containing 6 per cent of water (Column A).

(*b*) *Alumina deactivation*. Alumina in the columns was then deactivated to allow the desorption of the PAH. For this purpose, a stream of nitrogen, saturated with water, was passed through columns B and C for one hour in a downward direction. To retain the naphthalene eventually drawn off from alumina by the gaseous stream, a gas dispersion tube was attached at the bottom of column C and the fritted glass disc was immersed in the cyclohexane.

(*c*) *Chromatographic separation of PAH*. Following deactivation, column C (lighter compounds) was made to overlap a new column containing alumina with 4 per cent of water and the desorption of the PAH was carried out with cyclohexane. The PAH passed quantitatively in the lower column where the chromatographic development occurred. Column B (heavier compounds) was made to overlap another new column containing alumina with 5 per cent of water. The rest of the procedure was the same as that mentioned above.

(*d*) *Spectrophotometric analysis*. An automatic recording spectrophotometer was used for qualitative and quantitative analysis. The absorption spectra of the eluates were recorded and quantitative estimations were carried out by the base-line technique, according to the method of Cooper[6]. Cyclohexane was used for all determinations.

RESULTS

The analytical results are summarized in *Tables 1* and *2*, and illustrated in *Figure 4*. The recovered amounts of PAH were expressed as $\mu g/100$ m^3 of exhausts.

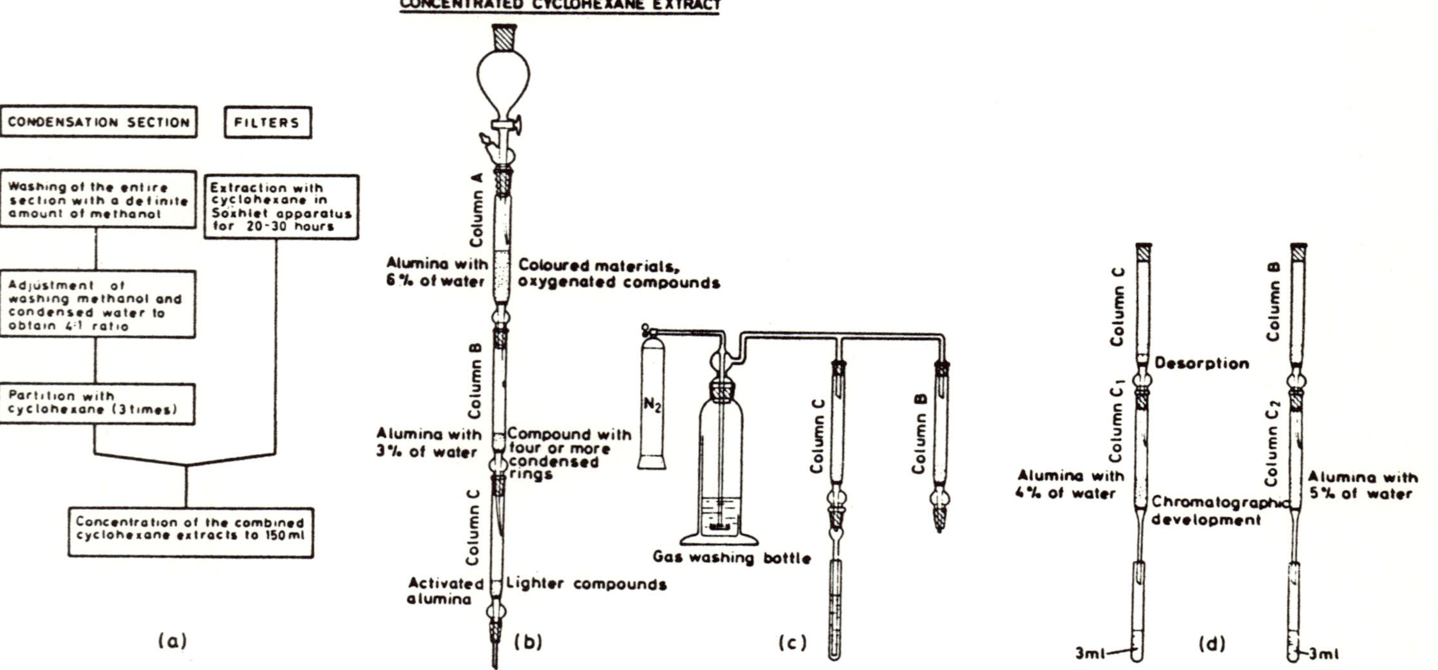

Figure 3. Schematic diagram of the analytical procedure.

146

Table 1. Comparison of polycyclic aromatic hydrocarbons from gasoline engine and liquefied petroleum gas engine exhausts. (µg/100 mc of exhausts)

Hydrocarbons	Gasoline	Liquef. gas	Variations%
naphthalene	140454·5	101449·2	− 27·7
acenaphthene	4000·0	—	
fluorene	4090·9	3840·6	− 6·1
phenanthrene	8500·0	3571·4	− 57·9
anthracene	1454·5	1242·2	− 14·6
pyrene	9081·8	6314·6	− 30·5
4-methylpyrene	500·0	217·4	− 56·5
fluoranthene	7436·3	5072·5	− 31·8
benzo (mno) fluoranthene†	2500·0	1138·7	− 54·5
1:2-benzanthracene	1963·6	993·7	− 49·4
chrysene	1272·7	683·2	− 46·3
perylene	545·4	283·1	− 56·3
3:4-benzpyrene	2000·0	1097·3	− 45·1
1:2-benzpyrene	3363·6	1086·9	− 67·6
3:4-benzfluoranthene	772·7	393·4	− 40·1
11:12-benzfluoranthene	1354·5	569·4	− 57·9
anthanthrene	454·5	434·8	− 4·3
1:12-benzperylene	3900·0	1956·5	− 49·8
coronene	1854·5	786·7	− 57·6

† Also 2:13-benzfluoranthene

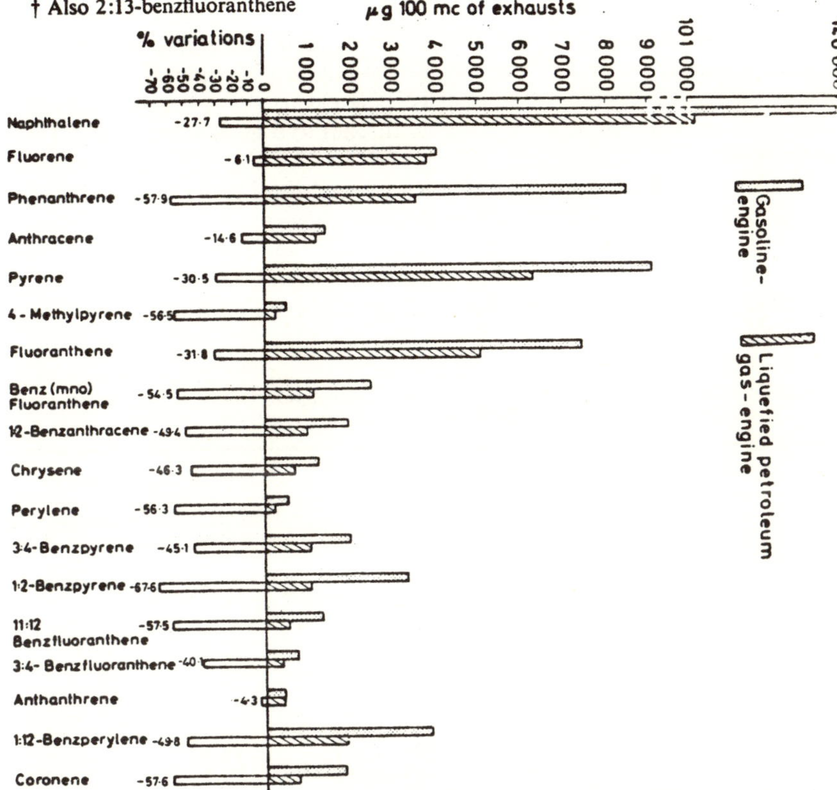

Figure 4. Comparison of aromatic polycyclic hydrocarbons from gasoline engine and liquefied petroleum gas engine exhausts

(i) In the gasoline engine exhausts, nineteen PAH were identified and determined: naphthalene, acenaphthene, fluorene, phenanthrene, anthracene, pyrene, 4-methylpyrene, fluoranthene, benzo(mno)fluoranthene, 1:2-benzanthracene, chrysene, perylene, 3:4-benzpyrene, 1:2-benzpyrene, 11:12-benzfluoranthene, 3:4-benzfluoranthene, anthanthrene, 1:12-benzperylene and coronene (*Table 1*). Among the detected compounds, some possess strong carcinogenic activity to animals under several experimental conditions, some show smaller activity and others are considered, although not unequivocally, inactive. Several other peaks of unidentified compounds were observed in the absorption spectra of the eluates.

(ii) In the liquefied petroleum gas engine exhausts, eighteen PAH were identified and determined: naphthalene, fluorene, phenanrtrene, anthracene, pyrene, 4-methylpyrene, fluoranthene, benzo (mno) fluoranthene, 1:2-benzanthracene, chrysene, perylene, 3:4-benzpyrene, 1:2-benzpyrene, 11:12-benzfluoranthene, 3:4-benzfluoranthene, anthanthrene, 1:12-benzperylene and coronene (*Table 1*). The absorption spectra of the collected fractions were almost identical with those observed for gasoline-engine exhausts. Acenaphthene seemed to be present but it could not be estimated quantitatively. In the trials in which the lift truck operated with a load on the simulated working schedule, the PAH concentrations in the exhausts proved to be notably inferior to those found in the gasoline engine operating under the same conditions. Percentage reductions (*Table 1*) varied, for individual hydrocarbons, from a minimum of 4·3 (anthanthrene) to a maximum of 67·6 (1:2-benzpyrene).

(iii) In the trials in which the engine powered by liquefied petroleum gas was idling at 500 rpm and at zero load, the concentrations of the individual compounds proved to be markedly inferior to those found for the same fuel in the trials based on the simulated working schedule. Naphthalene is an exception, whereas acenaphthene and 11:12-benzfluoranthene were not present (see *Table 2*).

Table 2. Polycyclic aromatic hydrocarbons from liquefied petroleum gas engine exhausts (μg/100 mc of exhausts).

Hydrocarbons	Work cycles	Idle	Variations %
naphthalene	101449·2	114075·0	+ 12·4
fluorene	3840·6	934·0	− 75·7
phenanthrene	3571·4	1313·0	− 63·2
anthracene	1242·2	178·6	− 85·6
pyrene	6314·6	1029·4	− 83·7
4-methylpyrene	217·4	31·5	− 85·5
fluoranthene	5072·5	735·3	− 85·1
benzo (mno) fluoranthene	1138·7	241·6	− 78·8
1:2-benzanthracene	993·7	168·1	− 83·0
chrysene	683·2	126·0	− 81·5
perylene	238·1	63·0	− 73·5
3:4-benzpyrene	1097·3	388·6	− 64·6
1:2-benzpyrene	1086·9	231·1	− 78·7
3:4-benzfluoranthene	393·4	94·5	− 75·6
11:12-benzfluoranthene	569·4		
anthanthrene	434·8	283·6	− 34·7
1:12-benzperylene	1956·5	693·3	− 64·5
coronene	786·7	273·1	− 65·3

Naphthalene is a highly volatile compound and it is therefore difficult to ascertain losses during sampling and subsequent treatments. In addition, according to Moore and Monkman[22], the presence of the parent substances is sufficient "to shift the diagnostic ultraviolet absorption peaks a few millimicrons, or enough to make positive identification doubtful". The same can be said of acenaphthene. With regard to 11:12-benzfluoranthene, however, it is possible that this hydrocarbon was present in amounts below the detection limits of the analytical method.

The marked reduction in PAH emissions found in the idling engine seems to disagree with the results of other authors who maintain that high production of all PAH occurs when the engine is idling at a lower number of revolutions per minute and decreases as the speed of the engine increases. One must remember, however, that these statements are based on trials in which the engine was running at a constant speed. In fact, evidence was presented that the production of PAH is greatly influenced by change of speed: the largest production of all PAH occurred during acceleration and deceleration These two conditions prevail while the lift truck is running on the simulated working schedule. Thus, the high quantities of PAH emitted in the exhausts are understandable.

(iv) The results of this study show that, from a qualitative point of view, the formation of PAH is independent of the fuel used. Thus the same compounds are found in the gasoline-engine and in the liquefied petroleum gas-engine exhausts. On the basis of several studies, a free radical mechanism appears to be the most logical way of explaining the presence of PAH in combustion products from all types of fuels. Kinney and Crowley[17], Egloff[8] Tebbens and co-workers[28], Badger and co-workers[1, 2] and other authors have suggested that the PAH are formed at high temperatures from aliphatic and simpler aromatic hydrocarbons by free radical reactions. These radicals recombine in various ways to give less hydrogenated and more condensed products.

An attempt by Tebbens and his co-workers[28] to determine the relationship of combustion to the formation of condensed-nuclei aromatic compounds has demonstrated the presence of multinuclear aromatics among combustion products. It has also been suggested that PAH in the exhaust products of internal combustion engines are produced by a mechanism involving acetylene as the starting material, this hydrocarbon resulting from pyrolytic breakdown of the aliphatic hydrocarbon fuels. At high temperatures, acetylene condenses through a free radical mechanism to form higher molecular weight aliphatic and aromatic hydrocarbons.

CONCLUSIONS

An engine exhaust sampling apparatus for PAH determination was constructed. The apparatus guarantees the quantitative recovery of organic materials of high molecular weight and the accurate measurement of the actual gas volume sampled. Exhaust gases were analyzed for PAH by an analytical procedure, previously described, which is a combination of chromatography (on alumina) and absorption spectrophotometry. A preliminary adsorption of PAH onto a series of superimposed alumina columns makes possible:

(a) the avoidance of evaporation to dryness or small volume and of the resulting losses due to the volatility of polynuclear compounds;

(b) the purification of the extracts from many extraneous materials which give rise to intense background in the ultraviolet absorption spectra;

(c) a first separation of PAH into two large fractions.

In the gasoline engine exhausts, nineteen PAH were identified and determined. Several other peaks for unidentified compounds were observed. In the liquefied petroleum gas engine exhausts, eighteen PAH were identified and determined. The absorption spectra of the collected fractions were almost identical with those observed for gasoline engine exhausts. The PAH concentrations all proved to be notably inferior. The results support the evidence that the formation of PAH, from a qualitative point of view, is independent of the fuel used. A free radical mechanism appears to be the most logical way of explaining the presence of PAH in combustion products from all types of fuels.

References

[1] G. M. Badger, R. W. Kimber and T. M. Spotswood. *Nature* **187**, 663–665 (1960).

[2] G. M. Badger. *Natl. Cancer Inst. Monograph* No. 9, 1–14 (1962).

[3] C. Bayley, A. Javes and J. Lock. Vth *World Petroleum Congress, Sect.* IV, Paper 13, N, York (1959).

[4] C. R. Begeman and J. M. Colucci. *Natl. Cancer Inst. Monograph No.* 9, 17–57 (1962).

[5] J. M. Colucci and C. R. Begeman. *J. Air Poll. Control Ass.* **15**, 113–122 (1965).

[6] R. L. Cooper. *The Analyst* **79**, 573–579 (1954).

[7] S. Copplestone. *New Zealand Med. J.* **61**, 97–99 (1962).

[8] G. Egloff. *Am. Chem. Soc. Monograph Series* 73, New York (1937).

[9] V. A. Gofmekler, M. D. Manita, D. Manusa, Zh. Zhants and L. L. Shepanov. *Gigiena i Sanit* **28**, 3–8 (1963).

[10] D. Hoffmann and E. L. Wynder. *Anal. Chem.* **32**, 295–297 (1960).

[11] D. Hoffmann and E. L. Wynder. *Natl. Cancer Inst. Monograph* No. 9, 91–110 (1962).

[12] D. Hoffmann and E. L. Wynder. *J. Air Poll. Control Ass.* **13**, 322–327 (1963).

[13] D. Hoffmann and E. L. Wynder. *Cancer* **15**, 93–102 (1962).

[14] Y. Kobayashi, T. Abe, H. Takeno, S. Kanno, S. Fukui and S. Naito. *Japan Analyst* **12**, 1057–1062 (1963).

[15] P. Kotin, H. L. Falk and M. Thomas. *A.M.A. Arch. Indust. Hyg.* **9**, 164–177 (1954).

[16] P. Kotin, H. L. Falk and M. Thomas. *A.M.A. Arch. Ind. Health* **11**, 113–120 (1955).

[17] R. E. Kinney and D. J. Crowley. *Indust. & Eng. Chem.* **46**, 258–264 (1954).

[18] M. J. Lyons. *Brit. J. Cancer* **13**, 126–131 (1959).

[19] M. J. Lyons. *Natl. Cancer Instit. Monograph No.* 9, 193–199 (1962).

[20] R. Lawrence, E. Elsevier and L. A. Ripperton. *J. Air Poll. Control Ass.* **14**, 126–129 (1966).

[21] S. Mittler and S. Nicholson. *Industr. Med.* **26**, 135–138 (1957).

[22] G. E. Moore, J. L. Monkman and M. Katz. *Natl. Cancer Instit. Monograph* No. 9, 158–169 (1962).

[23] E. Sawicki. *Natl. Cancer Instit. Monograph* No. 9, 201–220 (1962).

[24] E. Sawicki, J. E. Meeker and M. Morgan. *Arch. Environm. Health* **11**, 773–775 (1965).

[25] A. Scholl, J. Bubernak and R. Galford. *Carcinogenic effects of pyrolyzed hydrocabons from internal combustion engine exhaust fumes. A preliminary report. Proc. West Virginia Acad. Sc.* 1951.

[26] R. Stenburg, D. Von Lehmden and R. Hangebrauck. *Sample collection techniques for combustion source benzpyrene determination.* Presented at Am. Indust. Hyg. Assoc. Meeting 1961, Detroit, Mich.

[27] J. L. Sullivan and G. S. Cleary. *Brit. J. Ind. Med.* **21**, 117–123 (1964).

[28] D. Tebbens, J. F. Thomas and Mitsugi Mukai. *A.M.A. Arch. Indust. Health* **15**, 567–573 (1956).

[29] P. Valori. *Polycyclic aromatic hydrocarbons in polluted urban air*. Report presented at Congress on *Air Pollution as Public Health Problem*, Roma-Campidoglio 13–14 April, 1966.

[30] P. Valori, C. Melchiorri, A. Grella and G. Alimenti. *N. Ann. Igiene e Microb.* **17**, 311–324 (1966).

[31] P. Valori and A. Grella. *N. Ann. Igiene e Microbiol.* **17**, 351–382 (1966).

[32] P. Valori, A. Grella, C. Melchiorri and N. Vescia. *N. Ann. Igiene e Microbiol.* **17**, 383–414 (1966).

[33] E. L. Wynder and D. Hoffmann. *Cancer* **15**, 103–108 (1962).

CONTRIBUTION OF MOTOR VEHICLE EXHAUST, INDUSTRY, AND CIGARETTE SMOKING TO COMMUNITY CARBON MONOXIDE EXPOSURES

John R. Goldsmith, M.D.

INTRODUCTION

The exposure of community populations to possibly hazardous levels of carbon monoxide (CO) includes exposure to this substance in cigarette smoke, in connection with certain occupations, in households particularly from heating and cooking, and in the general community where motor vehicle exhaust is the major source. This paper contrasts these various forms of exposure with respect to three major attributes: (1) the proportion of people affected, (2) the magnitude of short- and long-term exposure in relation to the possibility of illness, and (3) the time-course of exposure in relation to the possible hazards.

CIGARETTE SMOKING

It has been known for a long time that cigarette smoke contains CO, but relatively few measurements have been reported. Dalham[1] has estimated that 54% of the CO in smoke is absorbed. Since CO is present in smoke in a far larger concentration than that of any other potentially hazardous ingredient, exposure to CO is proportionally greater than exposure to other active or potentially biologically active constituents such as nicotine and polynuclear hydrocarbons.

Although comparable absorption studies have not been reported for smoke from pipes and cigars, the extent to which elevated carboxyhemoglobin levels are observed in pipe and cigar smokers has quite consistently led to the conclusion that pipe and cigar smokers absorb less CO than do cigarette smokers.

The large amount of CO in cigarette smoke and the relative ease of evaluating its uptake lead naturally to considering CO uptake as in index of the uptake of other cigarette smoke constituents. Data such as that in FIGURE 1[2] give a fairly clear picture of the effects of (1) smoking cigarettes, pipes, and cigars, of (2) inhalation on the expired air CO, and, (3) by regression, on levels of carboxyhemoglobin. A relationship between carboxyhemoglobin and smoking has been reported from a variety of studies and only small deviations from the pattern in FIGURE 1 were observed in the material. However, two important deviations deserve emphasis.

Kjeldsen,[3] in studying atherosclerosis among a group of smokers and nonsmokers in Denmark, reported levels of carboxyhemoglobin in cigar smokers similar to those in cigarette smokers, but it must be emphasized that one of the occupational groups studied by Kjeldsen were workers in a cigar factory. It is tempting to assume that these individuals had an exceptional access to, and probably exceptional use, of cigars.

The other exceptional finding was the relationship between age and carboxyhemoglobin within smoking categories in a 1961 study of longshoremen. There is a fairly pronounced pattern of decrease in expired air CO with age (TABLES 1 and 2). The cause is not absolutely certain; it might reflect diminished uptake due to impairment of diffusing capacity. It has also been suggested that the decrease

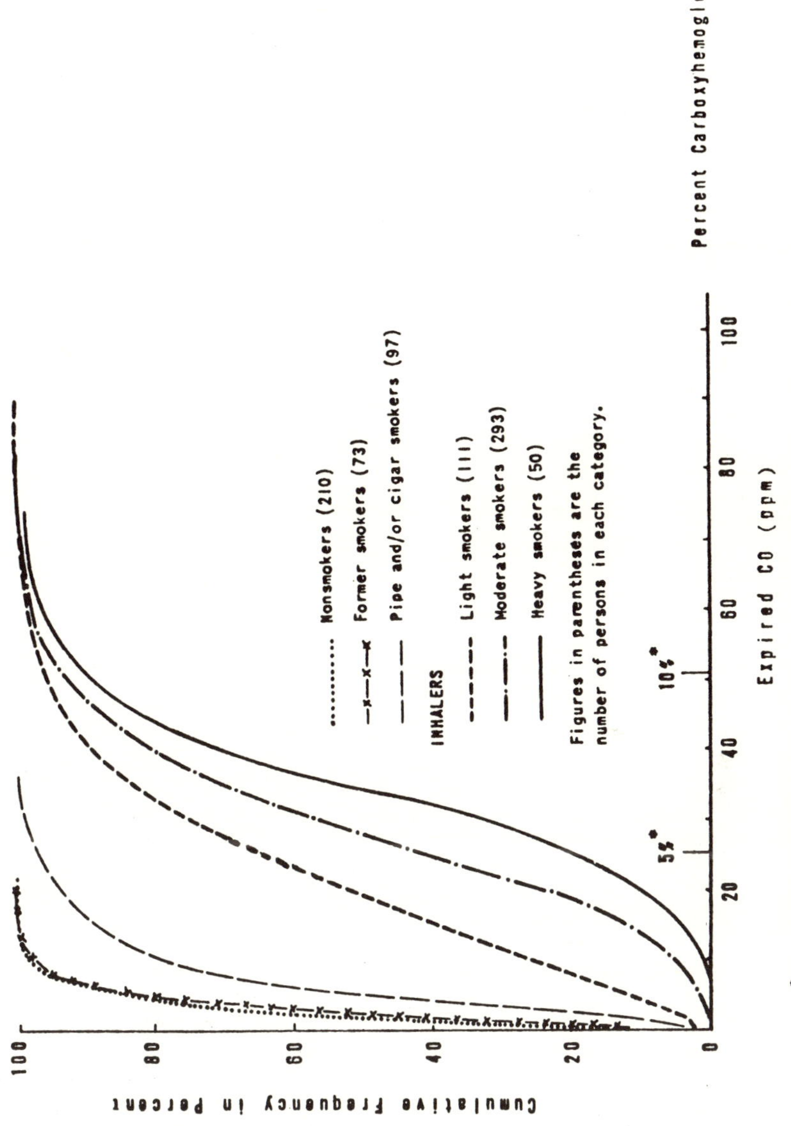

FIGURE 1. Distribution of expired CO in longshoremen, by smoking pattern, ILWU study, 1961 (normal subjects).

reflects the death of individuals who have absorbed high doses of CO and other constituents from cigarette smoke before they reach the sixth and seventh decades of life.

In the longshoremen study a subsample of the population had carboxyhemoglobin measurements made directly, and in three age groups the moderate smokers also showed a decrease of carboxyhemoglobin with age. The number of individuals is too small to show any trend of carboxyhemoglobin with age in the nonsmokers and the heavy smokers.

In a different population, one of outside telephone company workers studied in Los Angeles and San Francisco, a significant decrease in expired air CO with age is observed in both cities and a decrease, although not significant, is shown also for past smokers and pipe and cigar smokers, but there was no trend of expired air CO with age noted for nonsmokers.

MacIlvaine and colleagues[4] have studied the temporal variation of carboxyhemoglobin in both smokers and nonsmokers through the analysis of breath-holding expired air specimens, at two-hour intervals during waking hours on

TABLE 1

PROPORTION OF SMOKERS AND MEDIAN CONCENTRATIONS OF EXPIRED CO IN A
POPULATION OF LONGSHOREMEN

Category* 3,311 Individuals	Median concentration (parts per million) of CO measured in expired air	Median percentage of carboxyhemoglobin estimated from regression
Never smoked (23.1)	3.2	1.2
Exsmoker (12.1)	3.9	1.4
Pipe and/or cigar smoker only (13.4)	5.4	1.7
Cigarette smoker		
Light smoker (half pack or less) (13.0)		
Inhaler	17.1	3.8
Noninhaler	9.0	2.3
Moderate smoker (more than half pack or less than 2 packs) (31.3)		
Inhaler	27.5	5.9
Noninhaler	14.4	3.6
Heavy smoker (2 packs or more) (7.0)		
Inhaler	32.4	6.8
Noninhaler	25.2	5.6

* Values in parentheses are percentages of study population by smoking pattern.

TABLE 2

MEDIAN CONCENTRATIONS (IN PARTS PER MILLION) OF EXPIRED CO BY AGE OF SUBJECT
AND BY NUMBER OF CIGARETTES* SMOKED DAILY

Age of subject	Nonsmokers	Smokers		
		Fewer than 10 cigarettes per day	10–39 per day	>40 per day
<45	3.6	18.9	30.6	34.2
45–54	3.6	12.6	27.0	34.2
55–64	3.6	17.1	25.2	27.0
65–74	3.6	12.6	16.2	—
75–84	3.6	—	14.4	—

* Results obtained in the longshoreman study of TABLE 1.

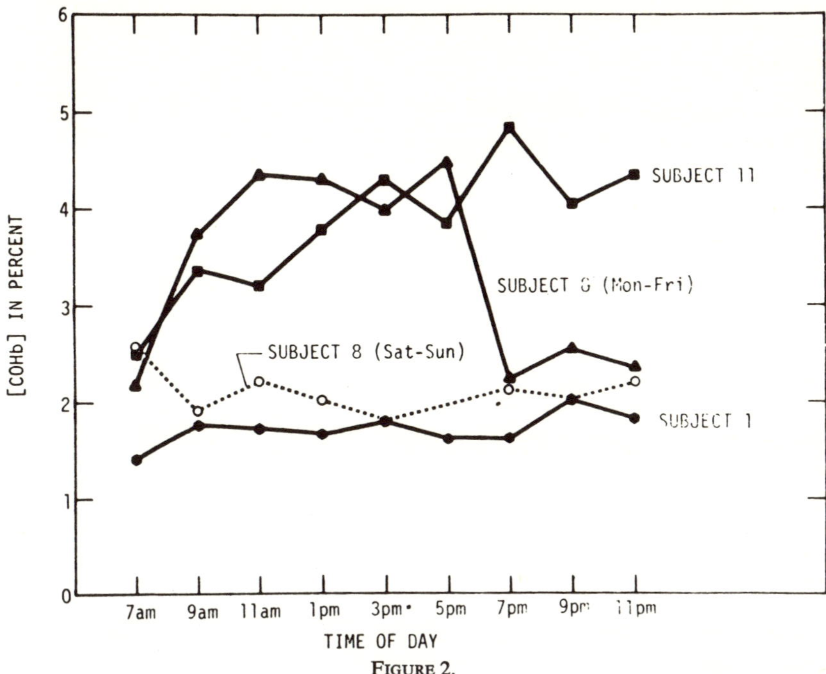

FIGURE 2.

seven consecutive days. Among their 15 subjects, only five were smokers. The nonsmoking subject with the highest level was employed in motor pool management and had occupational exposure to motor vehicle exhaust. The overall average levels obtained were 1.64% carboxyhemoglobin for nonsmokers (range 0.74 to 7.02) and 3.8% for cigarette smokers (range 1.09 to 3.05). The 7 AM specimen was generally lower than the specimens during the rest of the day; this was true for cigarette and nonsmokers. Selected data are shown in FIGURE 2. Excluding one cigarette-smoking subject with consistently low values, three out of the remaining four cigarette smokers had achieved by 9 AM essentially the carboxyhemoglobin level which they maintained during the rest of the day. This finding is mildly suggestive of the adjustment by cigarette smokers of the amount of inhalation and the maintenance of a relative state of equilibrium of carboxyhemoglobin for most of the day. In the recent studies from our Department there is suggestive evidence that in three out of seven cigarette smokers the carboxyhemoglobin level tended to increase from 11 AM to 5 PM, while in four the level either did not increase or decrease. Although this is flimsy evidence indeed, it is reason enough for asking whether the progressive increase of carboxyhemoglobin or the maintenance of an equilibrium value of CO is a characteristic which differs between different cigarette smokers. Such a pattern may be related to some of the effects of smoking.

Effects of the Type of Cigarette Smoking

In order to study the effects of different forms of smoking and of smoking different cigarette products, we studied seven cigarette smokers, two nonsmokers, and two pipe and cigar smokers through a schedule of "treatments" that consisted of (in the case of the cigarette smokers) not smoking at all for the day, smoking the usual brand of cigarettes, the use of the so called "smoke ring" through which the cigarette was passed with the usual brand of cigarettes, the smoking of a low nicotine cigarette, and of a nontobacco (lettuce) cigarette. In order to conform to the available office hours, the period between 9 and 11 AM was a period of smoking four cigarettes, one every half hour. The period between 11 AM and 5 PM was one in which the cigarette smoking frequency was optional, but the type of "treatment" was not. The estimation of CO uptake was made by the use of breath holding expired air, but blood samples for carboxyhemoglobin were taken from selected individuals at different times. The results are shown in FIGURES 3, 4, and 5.

FIGURE 3 shows that the 9 AM specimen, for each individual, tends to be within a rather narrow range. The subjects were requested not to smoke after midnight. The other pattern observed in FIGURE 3 is that within each subject the increase with different "treatments" is essentially the same. FIGURE 4 shows the behavior of expired air carbon monoxide when subjects' frequency of smoking was left to the choice of the subject. Although again there was no real contrast between the different forms of "treatment," three subjects had a sharp reduction in expired CO. Three subjects had, if anything, an inconsistent increase. One subject showed no change.

We also examined the relationship between the increase in carboxyhemoglobin during the 11AM to 5 PM period, the period in which the amount of smoking was optional, and the number of grams of tobacco consumed, determined by the difference between the weights of the fresh cigarettes and the cigarette butts. There is a highly significant linear regression which relates the data, although there is a substantial amount of scatter (FIGURE 5).

Impact of Cigarette Smoking

Cigarette smoking affects a very large number of persons in the community. Depending on the age and sex group, one may use 40–60% of adults as a rough estimate. The amount of exposure is from practically no carboxyhemoglobin up to that sufficient to produce 15% carboxyhemoglobin. The time course is inherently intermittent; there is a decrease each morning and an increase throughout the day. In some subjects there is evidence that the increase occurs early in the day and that subsequent smoking tends only to maintain the acquired level. Within smoking categories there is an apparent decrease in carboxyhemoglobin with age; the reasons for this are unknown. Pipe and cigar smokers in the United States have much less exposure to CO than do cigarette smokers.

OCCUPATIONAL EXPOSURE

Occupational exposure to CO is extremely common. It occurs in the metallurgical industries, with exposures to internal combustion engines in confined spaces, and with traffic control. Some risk of CO exposure occurs in many manufacturing processes, since it is readily formed under conditions of incomplete combustion. FIGURE 2 shows an example of carboxyhemoglobin estimates in a nonsmoker exposed in a parking garage. The variable patterns associated with occupational exposure are shown by data from a study of border crossing inspectors, carried out by our Department in cooperation with other organizations, and a study of exposures of fire fighters by Gordon.[5] I shall also present some data made available by Breysse[6] on exposures of cargo handlers on loading and unloading ships.

156

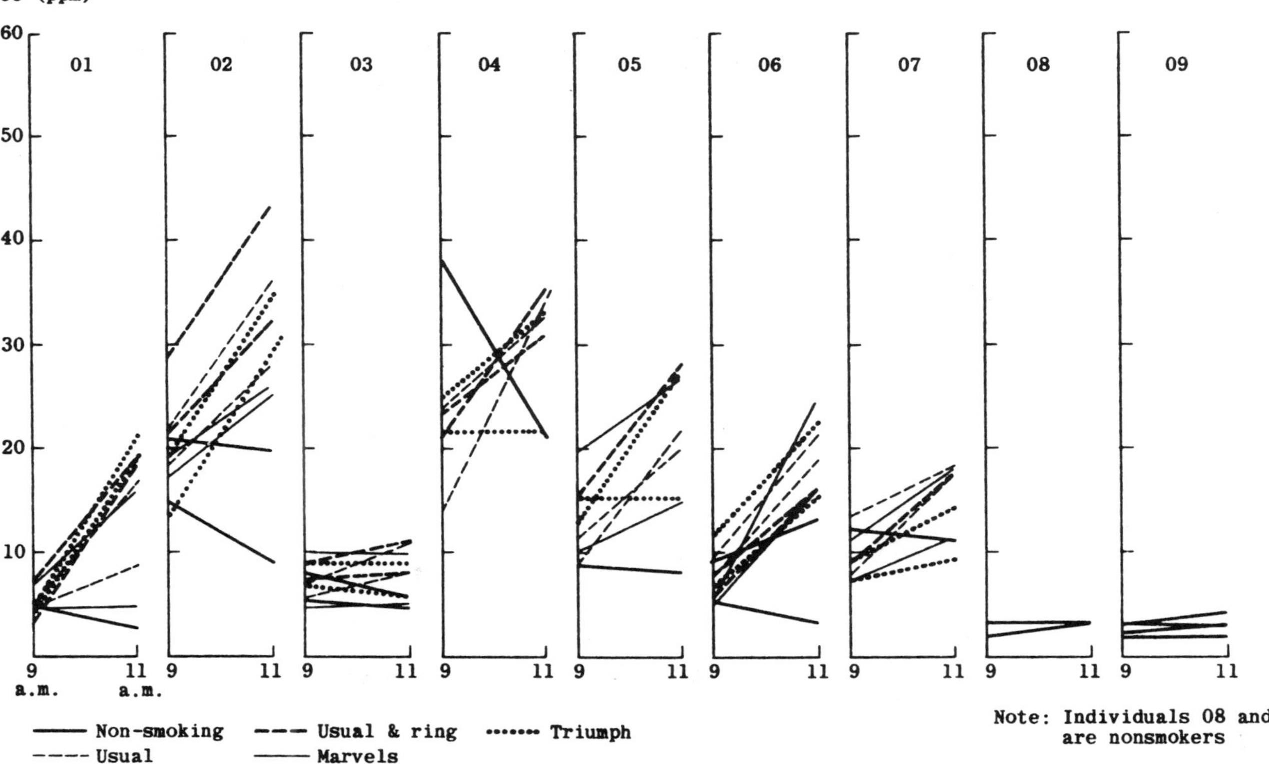

FIGURE 3. Expired air CO in smokers and nonsmokers.

157

We were asked to assist in evaluating the intensity of exposure to CO among a group of Public Health Service and other federal border-crossing inspectors working at San Ysidro, near San Diego, in the fall of 1969. The basic plan of the study was to obtain data on expired air samples before going on duty and after serving a tour of duty, to obtain a random sample of carboxyhemoglobin measurements in order to validate the relation between expired air CO and carboxyhemoglobin, and to estimate the atmospheric exposure and the meteorological conditions which were related to it at the time. At the time of this study there was an unusually severe amount of traffic congestion at the border-crossing stations, as a result of an intensified effort to control narcotics and dangerous drugs crossing the border; hence traffic delays were often as long as four to six hours. The cigarette smokers

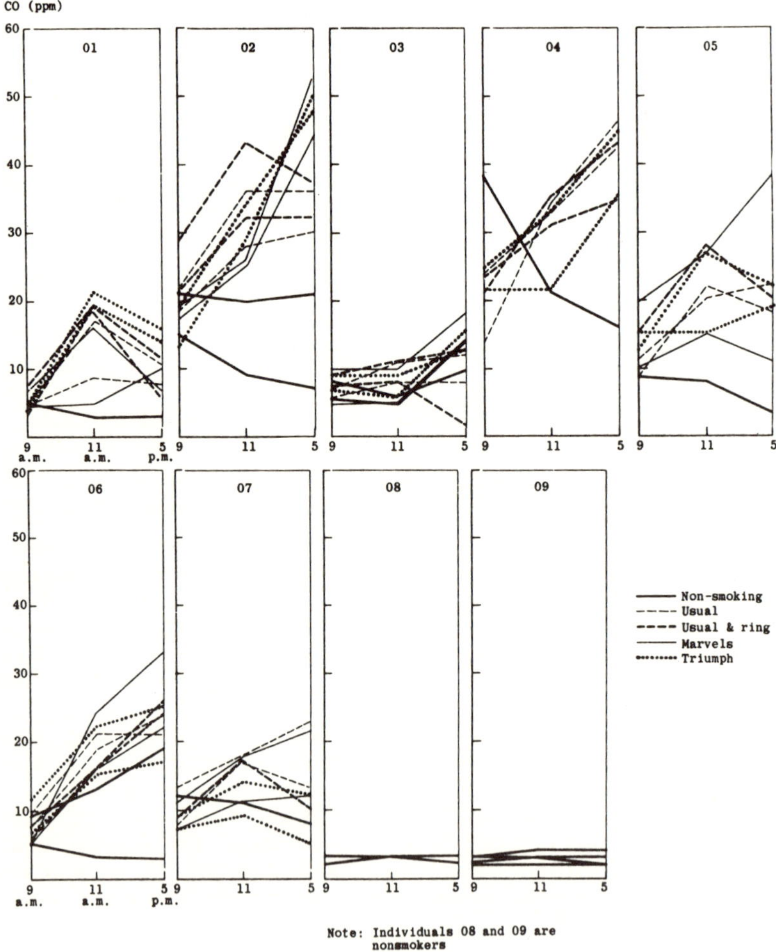

FIGURE 4.   Expired air CO in smokers and nonsmokers.

158

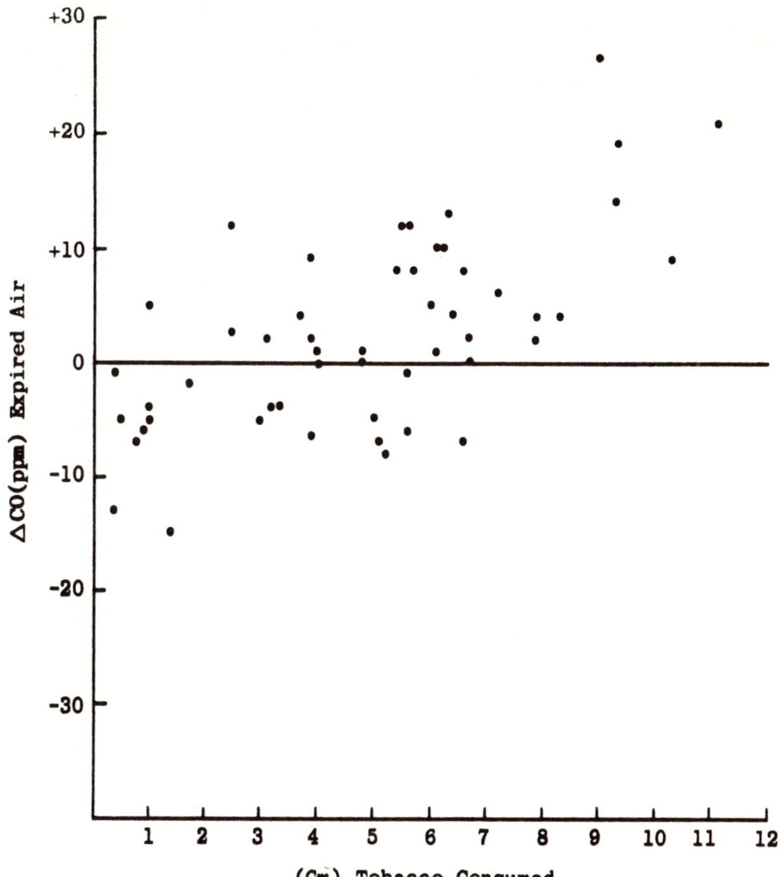

FIGURE 5.   ΔCO vs gm tobacco consumed, 11 a.m.–5 p.m. (all cigarette smokers).

were asked to refrain from smoking for eight hours prior to work and they were requested not to smoke during the period from before work to after work tests. As will be shown, these instructions were not always complied with. Data are shown in FIGURE 6 for three shifts, 8 AM to 4 PM, 4 PM to midnight, and midnight to 8 AM. During the first shift there was a wind which blew across the line of traffic, and either a decrease of expired air CO or no change during the shift. The latter was the most common in the nonsmokers and the former in the cigarette smokers. On the third shift the exposure was particularly intense as a result of a gentle wind in the line of the flow of traffic. Both smokers and nonsmokers showed a substantial increase in expired air CO. The second shift shows an intermediate effect. In general the nonsmokers showed an increase into ranges that were commonly observed in cigarette smokers without occupational exposure. The subjects who were not on the inspection line showed no increase or slight decreases, especially

159

if they were nonsmokers. When atmospheric conditions were favorable to producing exposure during the second and third shifts, there was a substantial exposure and relatively high carboxyhemoglobin levels were measured. The highest was in excess of 9% carboxyhemoglobin. In several cases in this small study the occupationally-related increase amounted to an equivalent of over 5% carboxyhemoglobin.

A number of other studies have reported the exposure of parking garage employees and policemen to motor vehicle exhaust, but in general the results have reflected a less intense type of exposure.

Gordon and Rogers[5] have recently reported on the carbon monoxide in fire fighters. All the measurements were made at a Denver Fire Station, which is perhaps significant, because the altitude at Denver with a fixed level of CO exposure would tend to make the resulting carboxyhemoglobin levels somewhat higher, due to the relative decrease in pO_2. Electrocardiograms were taken on a number of the subjects and a number of other biochemical indices were also obtained.

Over a 16-month period, exposure occurred in 31 fires of more than five minutes duration; 721 blood specimens and 686 electrocardiograms were taken from 35 subjects.

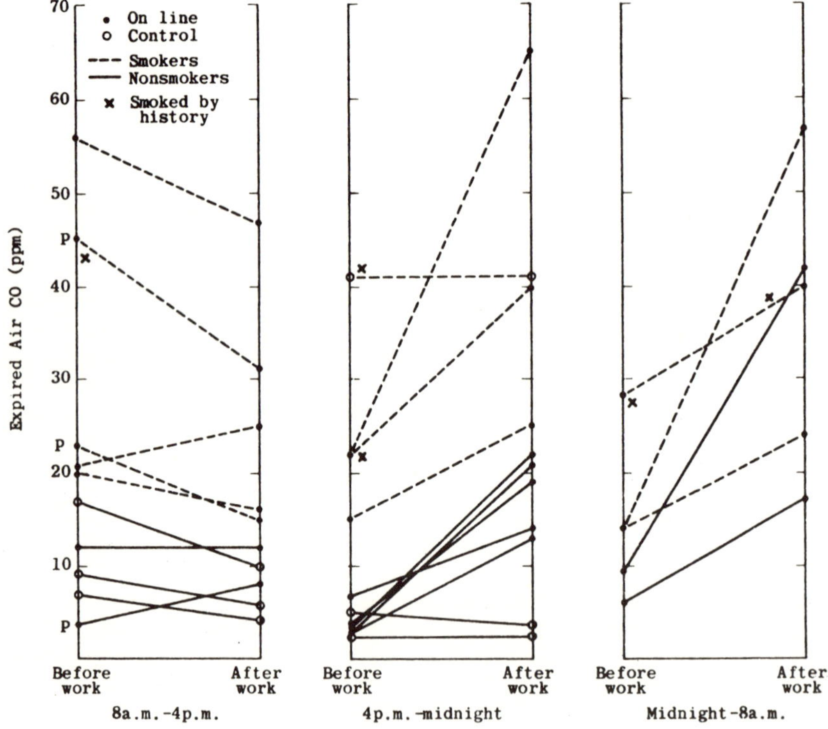

FIGURE 6. San Ysidro Border study.

This represenetd 221 individual exposures to fires. In four fires of five minutes or less duration, there were 13 subjects involved in some exposure. In addition there were six staged fires that permitted the evaluation of various respiratory apparatus used to protect against CO. One-half of the experimental subjects were smokers and all who smoked inhaled. Only 9 out of 138 baseline tests for carboxyhemoglobin showed a value in excess of 5%, the highest was 6.3%. After exposures to the longer fires, the median value was in excess of 10% carboxyhemoglobin and a tenth of the values were in excess of 25% carboxyhemoglobin. The highest observed value was 44% carboxyhemoglobin. The investigators reported that there appeared to be individual differences in the relative rates at which CO was cleared from the blood.

In 108, roughly one-half of the episodes, participation in the fire was associated with some changes in the electrocardiogram, but in some cases this was simply an increase in the cardiac rate. None of the electrographic changes persisted after 48 hours. In the studies of fires of brief duration, half of the carboxyhemoglobin values were 5% or greater.

Although the studies of Gordon and Rogers used and required the drawing of venous blood samples in order to make a measurement, Breysse[6] used the expired air sampling method and reports that 45 firemen representing exposure to five fires showed that 33 of the men had 5% CO hemoglobin or above, 15 were at 10% or more and two were at 20% or more. In Breysse's studies as well, the use of respirators was also protective.

Breysse also studied workmen loading cargo in the holds of ships using gasoline fork-lift trucks. Specimens were obtained before work, before lunch, after lunch, and after work. Selected subjects also provided blood specimens for testing the relationship between expired air and carboxyhemoglobin. In all, 92 nonsmokers and 108 smokers were studied. Among the nonsmokers all but 6% had before work carboxyhemoglobin levels of less than 3%, while nearly half the smokers had values in excess of 3%, seven of these were in excess of 7% carboxyhemoglobin. Sixteen individuals showed increases as a result of work of up to 10%, the highest was a 3-pack-a-day cigarette smoker who before lunch had up to a 20% carboxyhemoglobin level, but after work the level decreased to 13%.

IMPACT OF OCCUPATIONAL EXPOSURE

As is well known, fatalities from carbon monoxide exposure occur, so the maximal effect of occupational exposure of carbon monoxide is death. The usual effect of occupation is to cause increases of less than 10% carboxyhemoglobin. The usual pattern of work permits a substantial recovery if the work is intermittent. Estimates of the number of persons occupationally exposed in the United States are difficult to obtain, but at least three-quarters of a million probably have occasional occupational exposure.

HOUSEHOLD EXPOSURE

Each winter a number of people die as a result of CO poisoning associated with household heating, generally when it is improperly vented. The number of people affected to a lesser degree is virtually unknown. It seems unavoidable that a number of floor heaters produce CO, yet there are no modern surveys of this exposure. In primitive dwellings it has been shown that quite high exposures can occur Sofoluwe[7] and Cleary et al.[8]). This is obviously an area which needs a reevalua-

161

tion. Possibly the lack of attention to it is a reflection of the lack of attention generally by public health authorities to factors associated with housing. There is evidence from other countries that household heating is a factor of considerable magnitude in community exposure.

IMPACT OF HOUSEHOLD EXPOSURE

The major features of household exposure are that the subject is generally totally unaware of the risk, that the range of exposure can be extremely broad from less than 2% to that sufficient to cause fatality, that the exposure is often over a number of hours (often, tragically, the hours of sleep) and, that the exposure is inadequately evaluated.

COMMUNITY EXPOSURE TO MOTOR VEHICLE EXHAUST

The most abundant source of data concerning community exposure to motor vehicle exhaust is obtained from air monitoring, which indicates that at least in several of the metropolitan areas of the United States there have been a number of episodes in which exposures have been sufficient to produce an estimated 5% or more increase of the carboxyhemoglobin level in all of the exposed population. For example, in the CAMP air monitoring program (data between 1962 and 1967), Chicago, Denver, Philadelphia and Washington each exceeded an 8-hour, average of 30 parts CO per million, an amount expected to produce a 5%-carboxyhemoglobin level. In California, (the Sacramento and Los Angeles stations frequently exceed this; Sacramento twice in three years, (1962-65) and the Englewood station in Los Angeles 77 times. In interpreting this data for Los Angeles, it is necessary to note that the values may be excessively high because of the failure to correct for water vapor and the estimated excess may be from one to four parts per million. The reported average for daily exposure in Los Angeles is between 9.8 and 13 parts CO per million, an amount sufficient to produce, in light of the recent report of the National Academy of Sciences.[9] about a 2% increase in the carboxyhemoglobin level.

However no systematic studies have been undertaken to estimate the amount associated with community exposure, and therefore the monitoring system provides only tentative answers. The Los Angeles monitoring data represents sampling and sites thought to be representative of the general population exposure. As more and more people spend more and more time in automobiles on congested streets and freeways, there is an increased probability that individuals will be exposed to values greater than those of the general community exposure.

IMPACT OF COMMUNITY EXPOSURES

Tens of millions of people are involved in the community exposure to motor vehicle exhaust, resulting in an average blood level concentration of carboxyhemoglobin of less than 2%, but there could be exceptional circumstances leading to acute illness. For example, during extensive traffic delays associated with motor vehicle inspection and, occasionally, according to newspaper accounts, when a large number of cars attempt to leave parking facilities. The time course of exposure is somewhat unpredictable and this depends a great deal on meteorological conditions. In the winter these conditions are particularly conducive to a low mixing volume and hence to a substantial community exposure in California. In general, the exposures are elevated over periods of several hours sufficient to

TABLE 3

POSSIBLE CONTRIBUTIONS OF VARIOUS SOURCES OF CARBON MONOXIDE FOR THE COMMUNITY

	Number	Intensity	Time Course
Cigarette smoking	40–60% of adults	maxima of about 15% COHb*	intermittent diurnal patterns duration-decades
Occupation	? 750,000	usually $<$10% COHb high concentrations may cause death	8 hours or less per day, 5 days per week duration-variable
Household	? 100,000 no data	usually $<$2% COHb high concentrations may cause death	occasional, 4–8 hours may be repeated
Community air pollution (auto exhaust)	? 30,000,000 no data	usually $<$2% COHb high concentrations may cause illness?	unpredictable (4–8 hours) winter (in California)

* COHb = carboxyhemoglobin.

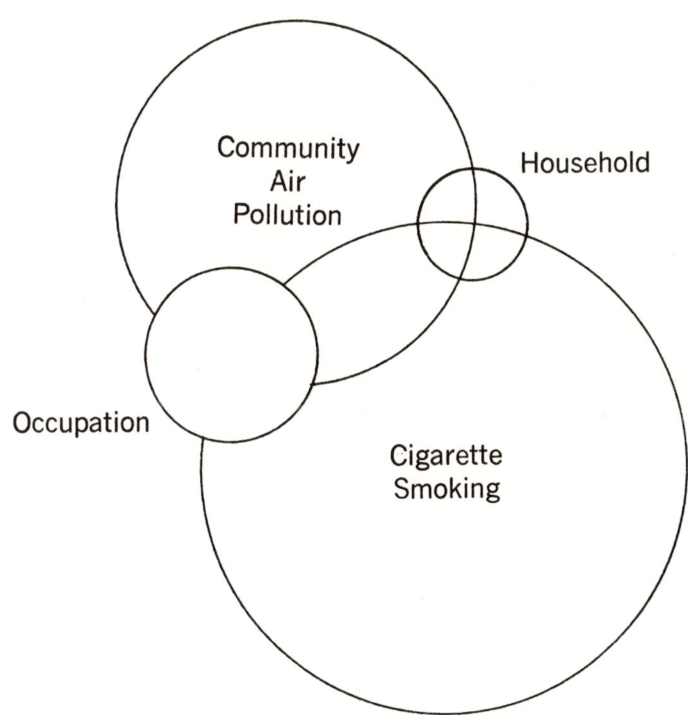

FIGURE 7.

163

produce an appreciable increment in carboxyhemoglobin. Detailed studies are certainly necessary in order to further estimate the severity and its importance.

DISCUSSION

The development of the breath-holding expired air method make it possible to do, with a reasonable effort, large-scale studies of selected populations to determine the magnitude of carboxyhemoglobin resulting from combined exposures. Obviously, persons exposed to multiple forms of CO exposure will have a greater carboxyhemoglobin level and presumably greater effects than those exposed to a single form alone. Whereas tens of millions of people are certainly exposed to at least one source of carbon monoxide exposure, hundreds of thousands are exposed to at least two sources, and perhaps many thousands to three sources of carbon monoxide. Because of the nature of the uptake and excretion processes in the lung, the effects of multiple exposures are probably somewhat less than additive.

SUMMARY

TABLE 3 and FIGURE 7 summarize the interaction of the various forms of exposure. Low-level carbon monoxide exposures are extremely widespread in the motorized, cigarette-smoking society of the United States.

REFERENCES

1. DALHAM, T., M. L. EDFORS & R. RYLANDER. 1968. Arch. Environ. Health **17**: 746.
2. GOLDSMITH, J. R., F. SCHUETTE & L. NOVICK. 1948. Proc. XIVth Intern. Congr. Occ. Health, Exc. Med. Intl. Congr. Ser. (no. 62).
3. KJELDSEN, K. 1969. Smoking and Atherosclerosis: The Role of Carbon Monoxide. Munksgaard.
4. McILVAINE, P. M., W. C. NELSON & D. BARTLETT. 1969. Arch. Environ. Health **19**: 83.
5. GORDON, G. S. & R. L. ROGER. 1969. A Report of Medical Findings of Project Monoxide. Redmond Memorial Fund, Intern. Assoc. Fire Fighters. Merkle Press, Washington, D. C.
6. BREYSSE, P. A. 1900. Use of Expired Air for Evaluating Carbon Monoxide Exposures. Multilith Rep. Dept. of Preventive Medicine, U. of Washington, Seattle, Wash.
7. SOFOLUWE, G. 1968. Arch. Environ. Health **16**: 670.
8. CLEARY, G. J. & C. R. B. BLACKBURN. 1968. Arch. Environ. Health **17**: 785.
9. NATIONAL ACADEMY OF SCIENCE-NATIONAL ACADEMY OF ENGINEERING. 1969. Effects of Chronic Exposure to Low Levels of Carbon Monoxide on Human Health, Behavior, and Performance. Washington, D. C.

Nitrogen Compounds other than NO_x in Automobile Exhaust Gas

H. P. Schuchmann and K. J. Laidler

Research on nitrogen compounds in automobile exhausts has so far been largely restricted to nitric oxide, NO, and nitrogen dioxide, NO_2, referred to collectively as NO_x; these oxides are present in amounts up to 3000 ppm.[1] One study,[2] however, has shown that from 1-6 ppm of NH_3 are present in the exhaust gas.

There have recently been carried out in this laboratory studies of the pyrolysis of ethane[3] and of acetaldehyde[4] in the presence of nitric oxide, at temperatures of 500°-600°C. In this work gas-chromatographic techniques were developed for the analysis and estimation of very small quantities of reaction products. The work revealed that a large number of nitrogenous products were formed by reaction between nitric oxide and these two substances, the main products being: ethane pyrolysis—acetonitrile (CH_3CN) and hydrogen cyanide (HCN); acetaldehyde pyrolysis—nitrogen (N_2), nitrous oxide (N_2O), hydrogen cyanide (HCN), methyl isocyanate (CH_3NCO) and ethyl isocyanate (C_2H_5NCO).

These results suggested to us the possibility that products such as these might be formed in the internal combustion engine, and be present in automobile exhausts, more particularly if the combusted mixture was not lean. The formation of such compounds under certain conditions might indeed explain the findings of Reid, Mingle, and Paul.[5] According to these workers, secondary air addition in the exhaust ports, when the engine is run on a rich mixture, leads to an increase in NO_x; with lean mixtures essentially no difference was observed with secondary air addition. Air addition may well cause a burning off of organic nitrogen compounds with the formation of NO_x.

Further evidence in the same direction is provided by an investigation of Starkman, Stewart, and Zwonow,[6] who used a spectroscopic technique to follow the concentration of NO during the initial expansion phase of the engine. It was shown that, with a 15% rich mixture, about one-third of the originally-formed NO subsequently disappeared. A 1% rich mixture showed a very slight NO disappearance, whereas a 9% lean mixture gave no NO decrease at all. This behavior might well be due to the scavenging of NO molecules by organic free radicals, which will be present in substantial amounts in the case of rich mixtures immediately after deflagration. Such scavenging will lead to organic nitrogen products, hitherto undetected in exhaust gases.

The analytical techniques[7] used in previous work are not suitable for the detection of organic compounds, for the following reasons:

1. In the process of drying the exhaust gas by passing it through cold traps, less volatile compounds would tend to be retained, especially if they were water soluble.

2. The wet methods used to determine NO_x are specific, being based on the colorimetric determination of an azo dye.

3. The optical methods used are based on the absorption of the exhaust gas at one selected frequency, rather than on the analysis of the entire spectrum.

4. Mass-spectrometric analysis[8] in the presence of residual hydrocarbons and partially oxidized species will give a complex spectrum, masking the presence of minor components.

The present study has employed techniques which are more suitable for the analysis of the expected organic nitrogen products.

Experimental

Exhaust gas from a 1963 Ford Anglia engine was collected under two conditions: slightly rich and rich. The gas chromatographic procedure was as described previously[4]; details of columns are given below.

1. Test for hydrogen cyanide

The exhaust gas was passed through a series of traps kept at 0° and −78°C, and the fraction which condensed at −196°C, consisting mostly of carbon dioxide, was investigated by gas chromatography. The columns employed were:

(a) Porapak Q 50/100 mesh, L = 3m, at 35°C
(b) 15% polypropylene glycol on Chromosorb ((AW-DMCS), 60/-70 mesh, L = 2.5 m, at 35°C

To circumvent excessive interference by the CO_2, whose retention time is much shorter than that of HCN, it was eliminated by passing the sample into either column (a) or (b), allowing the CO_2 to be eluted, and then reversing the flow and collecting at −196°C what had not yet emerged. The flow was then restored to its original direction and the collected substances warmed up and re-injected into either column (a) or (b). The analysis was complicated by the presence of a considerable number of minor components besides CO_2.

The analysis showed that the samples contained, in the CO_2, less than 10 ppm (by volume) of HCN; this is less than about 1 ppm in the total exhaust gas. However, there is some uncertainty, since it was observed that when known amounts of HCN were added to the sample a large part of this added HCN did not reappear in the analysis.

2. Test for nitrogen compounds condensable with water

The aqueous condensate was collected at the bottom of the exhaust condenser which cooled the entire exhaust to 10°C. After 1 ml of conc. H_2SO_4 had been added, 10 ml samples of the condensate were refluxed for 24 hours. Subsequent evaporation led to the destruction of the organic substance by the H_2SO_4. The clear residual solution was diluted and the NH_3 content determined by Nessler's method. In this way it was found that under rich conditions the condensate contained about 500 ppm (by weight) of NH_3, whereas the slightly rich mixture contained about 1500 ppm of NH_3. The stoichiometry of the combustion process can be approximately represented as follows,

$$6N_2 + \tfrac{3}{2}O_2 + \{CH_2\} \rightarrow 6N_2 + CO_2 + H_2O$$

air   fuel

so that water vapor composes about $^1/_8$ of the total exhaust gas. It follows that rich mixtures lead to the following results for the concentration of nitrogen-containing compounds other than NO_x in the total exhaust gas: rich mixtures—60 ppm (by volume); slightly rich mixtures—180 ppm.

With regard to the nature of these nitrogen-containing compounds, they were found to be almost completely involatile from acid solution (pH 4) and almost completely volatile from basic solution (pH 9). This eliminates acetonitrile (CH_3CN); a sample of this was subjected to the same procedures, where it was found to co-distil easily with water, being present in the distillate as well as the residue. It is concluded that NH_3 and lower amines (RNH_2) are produced in the internal combustion engine. It is noted that owing to the rear combustion temperature being lower the richer the mixture, less nitrogen will be fixed initially with the formation of NO, which explains the trend in the above result.

Discussion

The present work has indicated that HCN and CH_3CN are present in exhaust gases only in very small amounts. Our work on pyrolysis in the presence of nitric oxide has led to the conclusion that cyanides are formed by reaction of NO with organic free radicals by sequences such as the following:

$$CH_3 + NO \rightarrow CH_3NO \rightarrow CH_2{=}NOH$$
$$\rightarrow HCN + H_2O$$

$$C_2H_5 + NO \rightarrow C_2H_5NO \rightarrow$$
$$CH_3CH{=}NOH \rightarrow CH_3CN + H_2O$$

Organic radicals are present in substantial amounts in the combustion mixture, and such processes undoubtedly occur. It appears, however, that these cyanides are further converted into other products; by oxidation they will be converted back to NO, whereas under less oxidizing conditions they may be converted into NO, CO_2, and NH_3. The ammonia is probably formed by successive abstraction of H atoms by nitrogen-containing fragments.

Our result that amines and other organic nitrogen products may be formed in the combustion chamber is significant in connection with efforts to reduce NO_x in automobile exhausts. Chemical reduction of NO through addition of fuel into the exhaust system might not achieve reduction in the desired sense of causing it to form N_2, which is the only innocuous nitrogen compound; it probably merely converts NO_x into nitrogenous products such as amines which are equally objectionable.

In this connection, it is to be emphasized that any gas-phase process designed to convert NO_x into N_2 seems highly unpromising. In order for this conversion to occur two NO_x molecules must collide and react, an event which is very improbable at the low concentrations involved. Heterogeneous catalytic reduction of NO_x appears to offer a better prospect.[9]

Note added in proof

Since the completion of this study the identification by other workers of organic nitrogen compounds in car exhaust has come to our attention. Alkylamines have been isolated some time ago (A. J. Haagen-Smit, personal communication, 1971). Acetonitrile, methyl nitrate, acrylonitrile, and propionitrile have been detected by gas chromatography (T. A. Bellar and J. A. Sigsby, Jr., Environ. Sci. Technol. 4, 150 (1970)). Nitroolefins have been found as combustion products of hydrocarbons (J. Wasserberger et al., Ind. Med. Surg. 39, 225 (1970)), also by gas chromatography.

Acknowledgments

The authors extend thanks to Dr. E. G. Plett, J. Patry, K. Henry and J. Mar of the Department of Mechanical Engineering, Carleton University, Ottawa, for their cooperation in connection with the production of the samples of exhaust gas.

References

1. Hurn, A. W., "Mobile Combustion Sources," in *Air Pollution*, A. C. Stern, Ed., Vol. 3, New York, Academic Press, 1968.
2. Harkins, J. H. and Nicksic, S. W., "Ammonia in auto exhaust," *Environ. Sci. Technol.* 1, 751 (1967).
3. Esser, J. and Laidler, K. J., "The pyrolysis of ethane in the presence of nitric oxide," *Int. J. Chem. Kinetics*, 2, 37 (1970).
4. Schuchmann, H. P. and Laidler, K. J., "The pyrolysis of acetaldehyde in the presence of nitric oxide," *Int. J. Chem. Kinetics*, 2, 349 (1970).
5. Reid, R. S., Mingle, J. G., and Paul, W. H., "Oxides of Nitrogen from Air Added in Exhaust Ports," SAE Progress in Technology, Vol. 12 (Vehicle Emissions II), p. 230 (1966). SAE 660115.
6. Starkman, E. S., Stewart, H. E., and Zvonow, V. A., "An Investigation into the Formation and Modification of Emission Precursors," Int. Automotive Eng. Congr., Detroit, Mich., January 13–17, 1969. SAE 690020.
7. Dimitriades, B., "Methods for Determining Nitrogen Oxides in Automotive Exhausts," U.S. Bur. Mines Rept. Invest. 7133 (1968), 29 pp.
8. Campau, R. M. and Neerman, J. C., "Continuous mass spectrometric determination of nitric oxide in automobile exhaust," SAE Trans., 75 (1967). SAE 660116.
9. Starkman, E. S., "Eliminating exhaust CO and NO—It's possible," *SAE J.*, 28 (July 1969).

An Evaluation of the Fuel Factor
Through Direct Measurement of
Photochemical Reactivity of Emissions

Basil Dimitriades
B. H. Eccleston
R. W. Hurn

Findings in research at the Bureau of Mines' Bartlesville Petroleum Research Center show that photochemical reactivities of vehicular emissions are reliably measured in laboratory experiments in which smog manifestations are observed directly. Results of the direct smog-chamber measurements reveal that the photochemical behavior of emissions may differ significantly from the behavior that is predicted from the exhaust composition using reactivity scales. The concept of direct measurement of reactivity was applied to determine differences in characteristics of emissions from 20 passenger vehicles, each tested using 10 different fuels. The primary objective of the fuel study was to assess the over-all effect on vehicle emissions of fuel modifications designed to reduce the photochemical pollution associated with automotive evaporative losses. A similar, brief, comparative study of leaded and nonleaded fuels was also made. Reducing volatility was found to reduce the over-all smog potential of vehicle emissions but involved some penalty by way of increased exhaust emissions. Replacing light olefin with the corresponding paraffin also reduced over-all smog potential and in this case exhaust reactivity was not affected. In general greater smog potential was found to be associated with prototype nonleaded fuels than with leaded fuels typical of products currently marketed.

Hydrocarbons differ markedly in behavior as atmospheric contaminants; some react rapidly to form smog—some react slowly—some are virtually non-reactive. In automotive emissions, hydrocarbon species that contribute significantly to photochemical smog account for no more than one-half of the total mass of hydrocarbon emission. However, diminishing the total amount of hydrocarbon discharged to the atmosphere usually does not reduce the several species in equal ratio. Thus, to arrive at the optimum approach to smog abatement, it is important not only to take account of the amount of hydrocarbon discharged but also to assess the resultant effect in terms of photochemical or smog-forming consequence.

Methods for calculating the photochemical reactivity of hydrocarbon emissions have been proposed and are useful in providing directional information on the effects of fuel modifications. However, it would appear that direct measurement of the photochemical behavior of emissions would provide a superior, more reliable indication of the photochemical effect to be expected in the atmosphere; such an approach was chosen for use in recent fuel studies by the Bureau of Mines at its Bartlesville (Okla.) Petroleum Research Center.

Direct measurement of photochemical reactivity was used in studying three fuel factors: (1) The effect of fuel volatility on vehicle emissions, (2) the effect of fuel front-end composition, and (3) comparative evaluation of emissions from leaded and nonleaded fuels. In each of these studies it was required that the fuel change be assessed both on the basis of quantity of emissions and on the basis of the pollution effect predictable for the emissions. Moreover, these assessments were to consider both exhaust and fuel-system emissions, and the measurement of photochemical effect was to reflect the net consequence of the effects of fuel changes upon the separate sources.

The concept of photochemical reactivity measurement: In the context of this study photochemical reactivity of vehicular emissions is to be understood as an expression of the contribution that the emission makes to photochemical pollution. Regrettably, complex factors are involved in assessing this "contribution," and no one measurement or combination of measurements is likely to provide a fully satisfactory index. Nevertheless, useful indices are attainable. Such indices can be defined in terms of products that are normally associated with pollution; alternatively, a reactivity index can be defined in terms of a subjective response to the pollution itself, e.g., the eye-irritation response. Such subjective response measurements are presently considered to be complex and inexact; therefore, it was decided that reactivity indices based on instrumental measurements would be used in connection with the studies reported here.

The instrumental measurement itself poses dilemmas. There are several smog manifestations that are not necessarily interdependent; therefore, a hydrocarbon or a mixture of hydrocarbons may exhibit different reactivity manifestations that are not interdependent. The familiar diagram of Figure 1 is an elementary illustration of the various smog manifestations in terms of chemical changes that occur when hydrocarbon and nitrogen oxides in proper mixture with air are irradiated with sunlight. The changes that occur are: (1) NO is converted into NO_2, (2) hydrocarbon disappears, and (3) the reaction products that are smog constituents appear concurrently with other of the changes.

No one of these reaction manifestations provides a fully satisfactory basis for quantifying the objectionable pollution effects. However, all have been studied extensively and from the results of these studies there is evidence that the over-all level of activity in the photostimulated hydrocarbon/NO_x system is reflected in the pattern of NO_2 formation. Moreover, unlike other reaction products, NO_2 is formed in all hydrocarbon/NO_x systems and would appear to be a product that is reliably associated with objectionable photochemical pollution. Finally, in addition to its having broad significance in the hydrocarbon/NO_x reaction, the NO_2 can be measured with the greatest precision and reliability to be had for any of the several smog manifestations. For these reasons, experimental observation of NO_2 formation, more specifically, the rate of NO_2 formation (R_{NO_2}), provides the basis for photochemical reactivity measurement in the Bureau of Mines experimental work with automotive emissions.

In connection with this choice, note that the rate function—the rate at which NO is converted to NO_2—is related to the time-integrated effect which is yield. As illustrated in Figure 1, the time at which oxidant appears and the concentration of oxidant at any given time are directly related to the time required to reach peak NO_2 level. This time is a direct function of the NO_2 formation rate. A relation somewhat parallel to that of Figure 1, and having implications as discussed, would be expected in any time-limited system—such as a sunlit day—that builds and proceeds from a starting point.

The experimental work in which these concepts of reactivity measurement were applied was done by the Bureau of Mines in cooperation with the American Petroleum Institute. Technical assistance in the experimental program and in selecting fuels was provided by representatives of that organization. (See Acknowledgments at the end of this paper.)

Experimental

Vehicles and Fuels

The experiments to relate fuel composition, emissions, and photochemical behavior — the subject of this report — are discussed in other papers.[1-3] One should refer to those reports for details concerning broad experimental objectives, experiment design, and procedures used in making the emission measurements. The following is intended only as a resume of the salient features of the experimental program.

Twelve 1966 and four 1968 model automobiles were used in the studies of fuel volatility and of fuel front-end composition. Eight 1968 vehicles, including four of those just mentioned, were used in the study of prototype nonleaded fuels. All vehicles were selected from production-line vehicles. The particular engine and chassis configurations were selected to represent models significant in the auto population, but not necessarily to represent the auto population. Rather, they were selected to represent the most prominent engine chassis, and drive train combinations, with additional selections to represent engine or vehicle features that might be

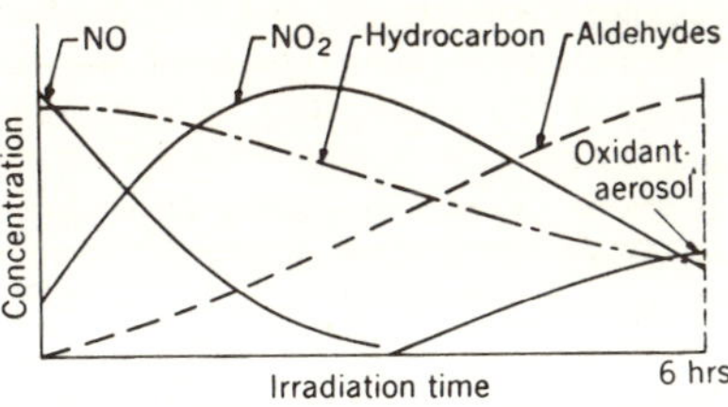

Figure 1. Reaction diagram of a hydrocarbon/NO_x/air/sunlight system.

of particular interest to the engineering study. A full range of vehicle sizes — heavy sedan to light compact — was represented in each of the fuel studies. Great care was exercised to determine that the vehicles were in typical operating condition and were maintained in such condition throughout the duration of the tests.

The base fuel for all of the tests was a 10-lb Reid vapor pressure (RVP), premium grade, leaded fuel that was formulated and prepared to represent an average U. S. summer gasoline. To study the effect of volatility, three additional gasolines were provided. These three additional gasolines were held as closely comparable in hydrocarbon composition as feasible and had Reid vapor pressures of 13, 7.9, and 5.3 lb, respectively. For the front-end composition study, two additional fuels with lower olefin contents were prepared from the base fuel. A quantity of this fuel was processed to remove, respectively, C_4–C_5 and C_4–C_7 hydrocarbons; these materials were then replaced with saturated hydrocarbons of comparable volatility. For comparison of emissions from leaded fuel with those from nonleaded fuel, the base fuel was paired with a nonleaded fuel of comparable octane and volatility. Composition of this latter fuel was modified to maintain octane quality while representing the typical fuel that would be manufactured if lead were to be removed from fuels generally. Three additional fuels were provided for the lead studies. These were a 10-lb, typical, or U. S. average, regular leaded fuel, a 10-lb, low-olefin,

171

Table I. Fuel properties.

Code No.	Grade	Lead, ml/gal	RVP	API Gravity	Distillation, °F % Evaporated			End Point	Aromatic[a]	Olefin[a]
					10	50	90			
Volatility Test Series										
1	Prem.	2.6	10.0	62.7	119	220	307	402	23	13
2	Prem.	2.4	12.9	64.3	106	220	310	398	21	12
3	Prem.	2.6	7.9	58.7	138	229	304	402	29	11
4	Prem.	2.4	5.3	56.7	157	235	307	400	28	12
Front-end Fuel Composition Tests										
1	Prem.	2.6	10.0	62.7	119	220	307	402	23	13
5	Prem.	3.1	10.1	62.8	120	221	308	398	23	9
6	Prem.	3.2	10.0	64.0	118	209	302	396	23	4
Leaded/Nonleaded Test Series										
1	Prem.	2.6	10.0	62.7	119	220	307	402	23	13
8	Prem.	0.0	11.0	52.9	117	238	315	370	50	8
7	Reg.	2.0	9.9	60.7	120	212	332	377	23	15
10	Reg.	0.0	10.4	56.5	119	208	309	354	43	8
11	Reg.	0.0	10.1	58.8	115	204	313	369	36	18

[a] Volume % by FIA analysis.

nonleaded prototype, and a 10-lb, high olefin, nonleaded prototype. Characteristics of the fuels are given in Table I.

Test Procedures

The objective of the experiments was to measure and characterize emissions from typical automobiles operated as they would be driven in typical city traffic. These conditions were approached in tests with the vehicles operated on a chassis dynamometer at ambient temperatures of 20°, 45°, 70°, and 95°F. The driving cycle on the dynamometer entailed a series of idle and cruise modes with accelerations and decelerations in a realistic pattern. The mode cycle was repeated 9 times to total $7\frac{1}{2}$ miles run in 21 min. Exhaust emissions were measured over the period of the driving cycle. Fuel-system losses were measured over the period of the driving cycle and for a 1-hr hot soak period thereafter. Neither diurnal nor tank-fill loss was considered.

Experimental data were taken as necessary to determine the mass of hydrocarbon emissions from tailpipe, carburetor, and fuel tank. Samples of these emissions were photoirradiated, and data from the subsequent reactions were taken as required to quantify the reactivity manifestation, which in this study was rate of NO_2 formation.

Sample Collection and Examination

Exhaust sample was collected in a Tedlar* bag that had been precharged with a volume of dry nitrogen 6 times that of the exhaust sample to be collected. A variable-rate proportional sampler[4] was used to control exhaust sampling rate to be a fixed fraction of the total exhaust flow. Sample from the collection bag (sampling having been completed) was used to charge the photoirradiation chamber and also to provide samples for chromatographic analysis and for measurements of total hydrocarbon, NO_x, CO, CO_2, H_2, O_2, and aldehydes. Evaporative emission samples — obtained in the form of liquid fuel condensates as explained below — were volatilized into clean, 34-liter stainless steel tanks and were mixed with nitrogen to a final total hydrocarbon concentration level of about 1000 ppmC. These pressurized mixtures were used to charge the chamber as well as to provide samples for the chromatographic analysis.

Fuel evaporated from the carburetor

* Polyvinyl fluoride (PVF) film manufactured by E. I. du Pont de Nemours & Co., Inc. Reference to a specific trade name is for identification only and does not imply endorsement by the Bureau of Mines.

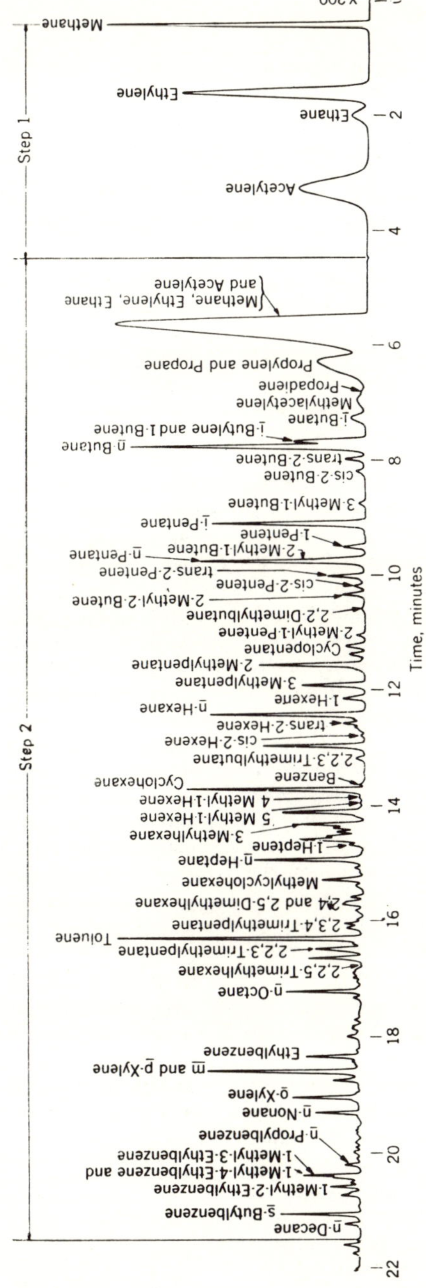

Figure 2. Chromatogram showing exhaust hydrocarbon separation.

173

could not be recovered for use as a reactivity test sample; alternatively, reactivity measurements were made upon samples of gasoline fractions obtained in a distillation designed to reproduce the characteristics of the evaporative loss. This was done as follows: Data from the vehicle loss measurements in all cases provided information that identified the loss as "X" percentage of the fuel in the tank or carburetor, respectively. (All carburetors were volumetrically calibrated so that this percentage loss value was convertible to mass; losses from the fuel tank were recovered and weighed.) Samples for reactivity assessment were, therefore, prepared in a single plate distillation to yield loss components at several, progressively higher degrees of fractionation. Specifically, fractions of 1, 2, 5, 10, 20, and 30% were distilled from each fuel for which reactivity assessments were to be made. Composition and other characteristics of these fractions were determined, and experimental reactivity data were obtained upon some fractions. From these data, curves were prepared for each fuel to show specific reactivity as a function of cumulative percent distilled. For any loss, expressed as percent of fuel evaporated, specific reactivity was then read from the prepared curve. This specific reactivity value was used, along with the weight of evaporative loss for any given test, to yield the value of the photochemical reactivity equivalent for the evaporative loss. (See "Experimental Measurement of Reactivity" that follows.)

Hydrocarbon composition data were obtained for all emission samples by chromatographic analysis. Briefly, the chromatographic method[5] utilizes a Perkin-Elmer Model 900 unit and two separatory columns. A packed column, operated isothermally, is used for separation of C_1–C_2 components; heavier hydrocarbons are separated in an open tubular column operating between $-100°$ and $+150°C$. A chromatogram of a typical exhaust sample contains approximately 200 peaks and is obtained in about 25 min (see Figure 2). Peak identities were established or approximated using (a) selective scrubber subtraction techniques, (b) retention data, and (c) information pertaining to stability of component hydrocarbons under dark and under irradiation (solar) conditions.

Experimental Measurement of Reactivity

The photochemical reactivity data were determined experimentally by irradiating the exhaust or evaporative emission sample in a 100-ft^3 chamber photoirradiated by artificial means. Design characteristics and operation of such an irradiation chamber have been reported.[6,7] Reactivity was measured and expressed in terms of rate of NO photooxidation (or NO_2 formation) in the chamber[7] (R_{NO_2}). Concentrations of hydrocarbon and NO_x in the initial chamber charge were adjusted to fixed values for each test. These conditions were: 3 ppmC total hydrocarbon plus 0.6 ppm NO_x (3:0.6 samples) for evaporative emissions and for exhausts from noncontrolled† automobiles; 3 ppmC total hydrocarbon plus 1.0 ppm NO_x (3:1 samples) were the initial concentrations for exhausts from controlled automobiles; the higher NO_x level in this latter case was unavoidable. In all cases the initial NO_x was proportioned about 90% NO and 10% NO_2. Resultant reactivity values for the 3:1 samples were adjusted to represent reactivities for 3:0.6 conditions using an adjustment factor of 0.95. This value was derived from independent exhaust reactivity studies, from which it was estimated that at a 3-ppmC total hydrocarbon level, increase of initial NO_x from 0.6 to 1.0 ppm causes an increase in rate of NO photooxidation of about 5%.

† Reference will be made to "controlled" and to "noncontrolled" automobiles in order to differentiate models designed with emission control features from those without this feature. Both 1968 regular production vehicles and 1966 models produced for sale in California are included in the "controlled" category.

The experimental values for NO photooxidation rate, from all samples, were converted to express the sample's reactivity in terms of its "specific ethylene reactivity equivalent," that is, in terms of a number that represents the sample's reactivity relative to that of an equimolar concentration of ethylene. For this purpose, every 2 weeks a standard mixture of 3 ppmC ethylene plus 0.6 ppm NO_x was irradiated in the chamber to provide regularly updated reference data. These data were plotted to reveal any systematic trend in the reference, and, by extrapolation or interpolation, to provide a reactivity value for ethylene applicable to any given test day.

The result from experimental measurement of an emission sample's reactivity is converted to its ethylene equivalent as follows: The value of ethylene R_{NO_2} for the test date is adjusted proportionately to represent reactivity of ethylene at a molar concentration (ppm) equal to that of the total hydrocarbon in the sample. The adjusted ethylene reactivity (R_{NO_2}) value is then divided into the comparable reactivity value for the sample and the resultant quotient was taken to express the sample's reactivity in terms of its ethylene equivalent, or, as defined, its "specific reactivity equivalent." Use of such "ethylene equivalent" units is desirable in that it allows comparison of reactivity measurement results in this study with those in other similar studies conducted under different experimental conditions. In addition, this unit guarantees that the reactivity value reference remains constant and is relatively independent of variable influences in the measurement system.

The term "photochemical reactivity equivalent" was chosen to express, i.e., quantify, the photochemical effect that is assignable to or predictable for a given amount of hydrocarbon emissions. This becomes a term that is intended to weight the amount of emission according to its photochemical activity. The unit of measurement is "grams ethylene"— this follows mathematically from expressing the specific reactivity of the emissions in terms of moles equivalent ethylene. The photochemical reactivity equivalent of a given weight of emissions is calculated as the weight of the emissions times its specific reactivity[‡] times the molecular weight of ethylene divided by the molecular weight of the hydrocarbon of the emissions.

In the discussion to follow, the simplified term "reactivity" may be used with reference either to "specific" reactivity or to the product term, photochemical reactivity equivalent. The authors have elected to use the simplified expression wherever the precise usage that is intended is apparent or a precise differentiation is not necessary to the sense of discussion.

Because experimental measurement of the reactivity of every exhaust and evaporative emission sample was impractical, a correlation method was developed and used to provide reactivity data for all samples generated in or applicable to the test program. By this method, about one out of four exhaust samples was experimentally measured for reactivity, and reactivities of the remainder were assigned using compositional data and reactivity/composition correlations established for those samples that were tested. Reactivities of evaporative emissions were determined in similar fashion. This program concerned with establishing the reactivity/composition correlations represents a distinct phase of the overall fuel study and will be referred to as the correlation study.

Photochemical reactivities of the emissions were also calculated from compositional data alone, using the linear summation technique described by Jackson.[8] By this method, reactivity of an emission sample is calculated as the weighted average of the individual hydrocarbon component reactivities; this is illustrated by equation 1.

[‡] Determined experimentally and expressed as moles equivalent ethylene.

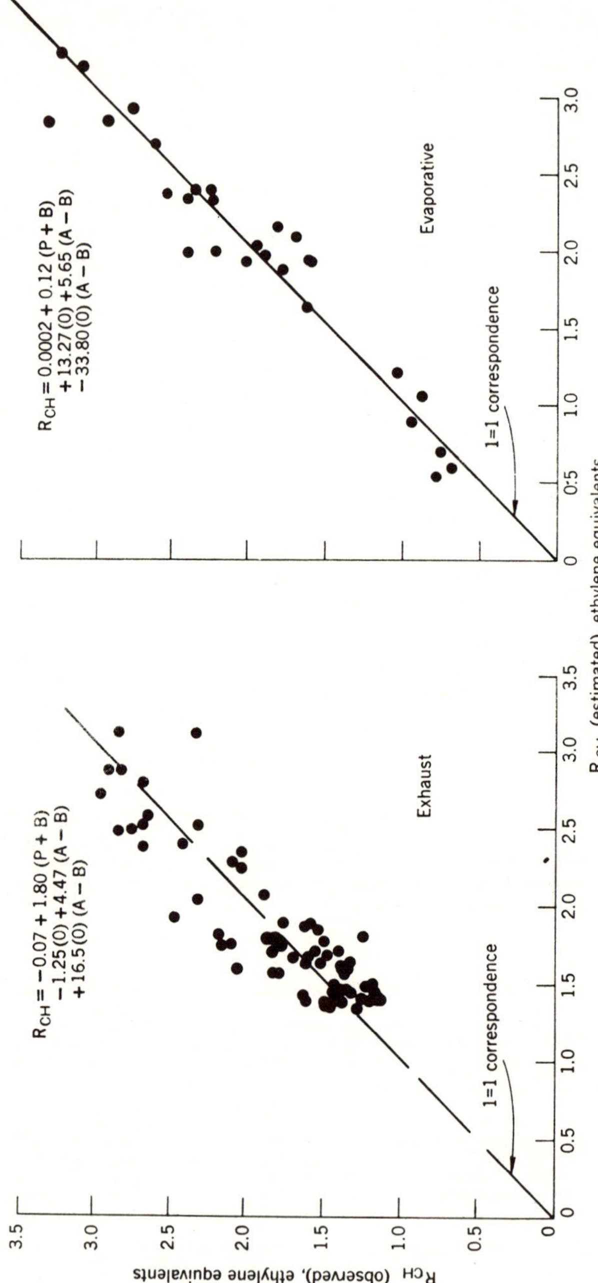

Figure 3. Reactivity/composition correlation for exhaust gas and evaporative losses.

$$R = x_1r_1 + x_2r_2 + \ldots x_ir_i + \ldots \quad (1)$$

The weighting factors are the components' mole fractions, x, as determined from chromatographic analysis; the component reactivity values, r_i, represent rate-of-NO-photooxidation reactivities and are those established by Glasson and Tuesday.[9]

Conclusions in this report regarding the relative effects of different fuel modifications are based wholly on rate-of-NO-photooxidation reactivity data obtained through direct experimental measurement. Linear summation reactivity data were generated and used in parallel, only for the purpose of demonstrating the extent to which the method of reactivity measurement may influence judgment upon the relative effectiveness of different fuel modifications in reducing levels of emission reactivity.

Eye-irritating reactivities of emissions were also computed from compositional data, using equation 1 and component reactivities representing eye-irritation potentials of the hydrocarbon components. Sources of such component reactivity data used in this study are discussed in the Appendix.

Results and Discussion

Results are reported and discussed separately for the two phases of the fuel study—(1) the reactivity/composition correlation study and (2) evaluation of fuel factors by emission reactivity measurement.

Correlation Study

The approach taken in the correlation study is based on development of a correlation equation in which correlation variables are chamber reactivity, R_{CH}, and compositional factors as illustrated by equation 2.

$$R_{CH} = K_0 + K_1[P + B] + K_2[O] +$$
$$K_3[A - B] + K_4[O][A - B] \quad (2)$$

where,

[ ]: mole fraction
P, B, O, A: paraffins, benzene, olefins, aromatics

The appropriate coefficients were established from the data of the correlation study in the following manner. For each sample of the correlation study the sum of the five compositional terms of equation 2 were equated to the sample's chamber reactivity, resulting in an equation with the five coefficients as unknowns. Such equations obtained for all samples of the correlation study were then treated statistically to render numerical values for the K coefficients. The resultant correlation equations were used to estimate chamber reactivity directly from compositional data. The extent to which the estimated values agree with those that were experimentally measured (R_{CH}), is shown graphically in Figure 3.

The correlation between chamber reactivities, R_{CH}, and reactivities computed by Jackson's linear summation method, R_Σ, is shown graphically in Figure 4.

The reactivity values that are shown are expressed in terms of ethylene equivalents. Linear summation reactivities were also expressed in terms of ethylene equivalents; this was done simply by using specific reactivity values for the sample components divided by the corresponding value for ethylene. In analyzing the correlation data it was assumed that when no hydrocarbon sample was injected in the chamber, both correlation variables had zero values, resulting in a (0,0) point in the correlation diagram. In actuality, this is not exactly true because chamber air alone does exhibit reactivity, the so-called background reactivity. In "ethylene equivalent" units, this background reactivity cannot be assigned a specific value, with any accuracy. However, such a correlation point must be used to justify interpolation of the correlation line in the region in which reactivity values approach zero. Because of this latter requirement, and because the background reactivity—in terms of NO photooxidation rate—was less than 20% of that of the least reactive sample in this study, it was felt that the assumption made was an alternative preferable to leaving out the (0,0) point.

Discussion of the results from the correlation study is centered around the appropriateness of the correlation equations of Figure 3, in predicting or providing estimates of chamber reactivity directly from compositional data. This direct method for prediction or estimating chamber reactivities will be referred to as the "Bureau of Mines (BuMines) correlation method." The appropriateness of the BuMines correlation method is illustrated best by comparing it with the only existing alternative method for predicting reactivity from composition, namely, the linear summation method. Because the direct measurement[6] of photochemical reactivity and indirect calculation by linear summation yield different values, the question is asked, "Which more faithfully reflects the sample's air polluting quality?". In the context of this question an examination of the correspondence of one to the other is useful.

Examination of correlation results shown in Figure 4 leads to two observations: (1) Linear summation reactivity values, R_Σ, correlate excellently with observed chamber reactivity values, R_{CH}. However, there is neither a 1:1 nor a simple linear correspondence for most samples. (2) Patterns of such R_Σ-versus-R_{CH} correlations are different for different sample types (see Figure 4). These observations point out the basic inadequacies of the linear summation method, namely: *A linear summation reactivity value in itself does not adequately reflect the reactivity of a hydrocarbon mixture in the hydrocarbon/NO_x system. Even more important, the inadequacy of the linear summation method varies with sample composition.* These results are reasonable when one considers the validity of the theoretical principles and assumptions upon which the linear summation method is based. These assumptions are: (1) Hydrocarbon components in the emission sample are correctly identified and measured. (2) Specific reactivities of hydrocarbon components have constant values, independent of hydrocarbon level, hydrocarbon composition, and individual hydrocarbon-to-NO_x ratio. Both of these assumptions are questionable. The identity of reactive components in automotive emissions, especially of those in exhaust, can be established only with limited accuracy using currently available routine methods. Specific reactivities are known to be affected by concentration when hydrocarbon concentration exceeds 1 ppm, and to all probability by hydrocarbon interaction also. In the present study, hydrocarbon level in the chamber was below 1 ppm at all times; however, hydrocarbon interaction and hydrocarbon-to-NO_x ratio effects have probably introduced reactivity/composition interactions that may not extend unperturbed to the atmospheric system.

In contrast to the linear summation method, the BuMines correlation method is not affected by uncertainties such as those discussed in the preceding paragraph. Validity of the method does not have a critical dependence on the accuracy with which hydrocarbon components in the sample have been identified. Accuracy of component identification is needed by the method only to the extent necessary to provide a rational classification of components in groups, e.g., paraffins plus benzene, olefins, and aromatics minus benzene. *Misidentification will not introduce error into the reactivity estimate as long as the same analytical method is used in the derivation and in the application of the correlation equations.* By the same reasoning, hydrocarbon interaction and NO_x effects are not expected to affect seriously the method's validity inasmuch as such effects are incorporated in the calibration of the method, that is, in the derivation of the correlation equation. Finally, reactivity values by the BuMines correlation method do show 1:1 correspondence with observed chamber reactivities; this is a natural result since the correlation method was de-

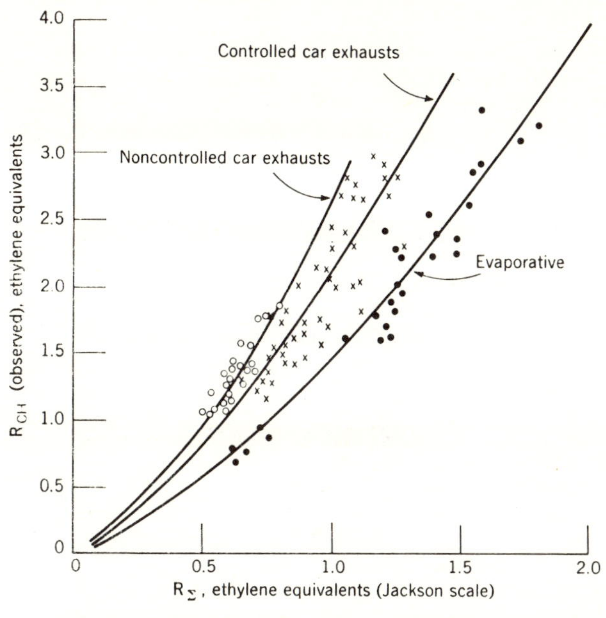

Controlled car exhausts: $R_{CH} = -0.04 + 1.32\, R_\Sigma + 0.82\, R_\Sigma^2$

Noncontrolled car exhausts: $R_{CH} = 0.01 + 1.41\, R_\Sigma + 1.12\, R_\Sigma^2$

Evaporative: $R_{CH} = -0.08 + 1.07\, R_\Sigma - 0.54\, R_\Sigma^2$

Figure 4. Reactivity/composition correlations for exhaust and evaporative emissions, Jackson scale. (These data are based on results from tests simulating a national average city trip plus evaporative losses during the subsequent one-hour soak. Data were taken from sixteen vehicles and the average of results from 70° and 95° F ambient test temperatures.)

[1] Reid vapor pressure.
[2] By Bureau of Mines chamber reactivity measurement.

signed so as to provide such correspondence.

While the BuMines correlation method was successful in meeting the needs of the fuel study, it is important that the limitations of the method be recognized in order to avoid misuse. First, the method is applicable only in studies in which experimental procedures for measurement of reactivity and of composition are the same as those used in the development of the method. Second, correlation equations can be used as predictor equations only upon samples similar in composition to those that were used to generate the correlation data. Finally, it is important to stress that a BuMines correlation equation does not constitute and should not be taken as the mathematical expression of a physicochemical entity, namely, the dependency of reactivity on composition; rather it is a mathematical expression of a statistical finding regarding correspondence of reactivity and composition values within a specified family of samples. Further, because in these correlation equations, component concentrations are expressed in terms of mole fractions, it follows that a change in any components' concentration level will automatically affect the mole fractions of all components in the mixture. Therefore, the value of an K coefficient in equation 2 does not reflect the extent to which a concentration change for the respective component (or group of components) affects the total mixture reac-

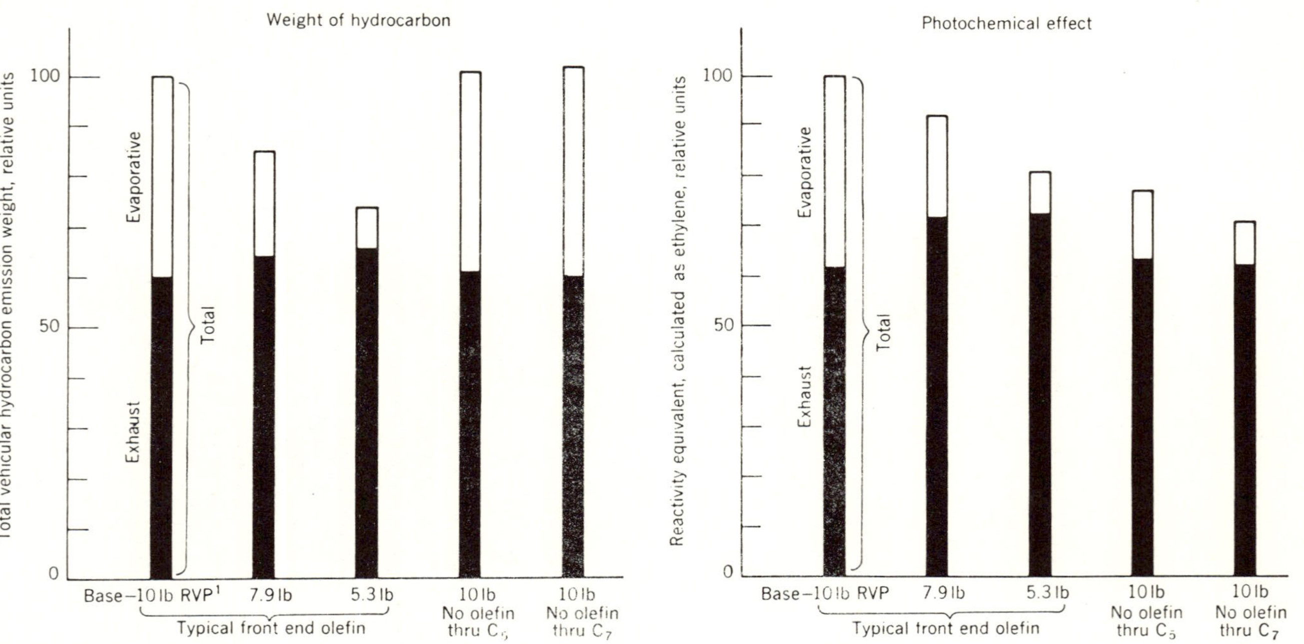

Figure 5. Dependence of weight and reactivity of hydrocarbon emissions on fuel front-end characteristics.

tivity. Rather it reflects the overall disturbance in component distribution that results from a change in one component's concentration. These comments are made to clarify the point that the significance of the K coefficients in the BuMines correlation equation is limited to that of coefficients in a correlation equation rather than of factors representing a chemical property of the respective components. In this respect the K coefficients differ from specific reactivities and may assume values that are not comparable with those of specific reactivities.

An alternative correlation technique, also for predicting reactivity from composition, was developed using the correlation equations that relate R_{CH} and R_Σ (Figure 4). The difference between this technique and the correlation method already discussed lies in the fact that here the compositional data are not used directly, but after manipulation (described by equation 1) that results in an R_Σ value. On these bases, this latter technique has been termed indirect correlation method, in distinction from the direct correlation method that was discussed in the preceding paragraphs. While both the direct and indirect methods have the same advantages that were shown to result from their correlative nature, a question may arise regarding the relative precision with which the two methods can predict chamber reactivity. Relative precisions of these two methods were measured and expressed in terms of amount of scatter present in the respective correlation diagrams. Such precision or scatter data are given in Table II. The data of Table II suggest that chamber reactivity can be predicted equally well by the two correlation methods. All in all the two correlation methods appear to be equally successful in predicting chamber reactivity from composition of hydrocarbon material in the automotive emissions. However, the direct method represents a far more promising and practical approach in that it may provide a useful predictor equation using less detailed, more readily obtained compositional

Table II. Precision(s) with which chamber reactivity, R_{CH}, is estimated by the direct and indirect correlation methods.

Estimation method	Standard deviation[a]	
	Exhaust emissions	Evaporative emissions
Direct correlation of R_{CH} with compositional data	0.26	0.24
Indirect correlation of R_{CH} with compositional data (R_Σ)	0.27	0.23

[a] Designates standard deviation of difference [R_{CH}(observed) − R_{CH}(correlation)] from zero.

data.

In conclusion, findings from the correlation data study were:

1. There is a well defined correlation between chamber reactivity and the hydrocarbon composition of an emission sample. Such correlations, established separately for each class of like type emissions, can be used to estimate reactivity from compositional data.
2. Reactivity estimates obtained through linear summation of compositional data — without making experimental reactivity measurements — do not agree with those based on direct experimental measurement or those obtained by the correlation methods developed in this work.

Influence of Fuel Composition on Vehicular Emissions

In this treatment of the data, it is convenient to discuss the two fuel modifications, fuel volatility and front-end fuel composition, in one section and to discuss separately the effect of formulating fuels without using lead to provide octane improvement. This arrangement of the report follows because the volatility and front-end fuel modifications are both relevant to evaporative emission control. They may be used as alternative measures to abate pollution from evaporative losses or they may be used in combination. For this reason, it is both convenient and useful to treat the two together. On the other hand the fuel modification that would be

Table III. Comparison of emissions from fuels of volatility series and modified composition series.

Code No.	RVP	Olefin Characteristic	Average of 70° and 95°F Tests[a]			
			Exhaust	Evaporative[b]	Total	Total as % of Base
			Gross Hydrocarbon Emissions, g per test			
1	10	Normal	46.3	30.7	77.0	100.0
3	7.9	Normal	49.6	15.9	65.5	85.1
4	5.3	Normal	50.4	6.3	56.7	73.6
5	10	None thru C_5	47.4	30.6	78.0	101.3
6	10	None thru C_7	46.8	32.1	78.9	102.4
			Hydrocarbon Emissions—Reactivity Equivalent Ethylene, g per test			
			Bureau of Mines Chamber Evaluation			
1	10	Normal	36.7	22.7	59.4	100.0
3	7.9	Normal	43.3	11.6	54.9	92.5
4	5.3	Normal	43.4	4.5	48.0	80.8
5	10	None thru C_5	37.1	8.6	45.7	76.9
6	10	None thru C_7	36.7	5.5	42.2	71.0
			GM NO_2 Formation Scale by Jackson			
1	10	Normal	18.4	13.7	32.1	100.0
3	7.9	Normal	18.8	7.0	25.8	80.3
4	5.3	Normal	18.9	2.8	21.7	67.4
5	10	None thru C_5	17.7	7.7	25.4	79.0
6	10	None thru C_7	16.9	6.8	23.7	73.8

[a] Each fuel was used in replicate tests on 16 vehicles at each of the two temperatures.
[b] Evaporative loss average of 70° and 95° F includes the logarithmic average of tank losses at 70° and 95°F. The average shown will not agree with a simple average of the evaporative losses shown for 70° and 95°F.

Table IV. Summary of emission data[a] for comparison of base and modified fuels. (All comparisons based on the average of results obtained at 70° and 95°F test ambients.)

	Fuel Number				
	1	3	4	5	6
RVP	10	7.9	5.3	10	10
Olefin, vol. %, FIA	12.8	11.5	12.1	8.9	3.8
Olefin distribution	Normal	Normal	Normal	None thru C_5	None thru C_7
Reactivity equivalent,[b] ethylene, relative units	100	92.5	80.8	76.9	71.0
Weight of hydrocarbon, (reactivity not considered), relative units	100	85.1	73.6	101.3	102.4

[a] The data are for the total of exhaust and evaporative losses measured during dynamometer tests to simulate combined typical city driving and hot-soak cycles.
[b] Bureau of Mines chamber reactivity measurement.

required to formulate fuels with adequate octane quality but free of lead additive presents an entirely different problem in fuel technology.

Effect of fuel volatility and front-end fuel composition. Data that are relevant to assessments of fuel modifications on the basis of levels and reactivities of emissions are given in Table III. These data are summarized in Table IV and illustrated in Figure 5. In Figure 5 it is shown that if the quantity of emissions is considered without regard to the photochemical effect, vapor pressure exerts a marked influence that acts primarily upon the evaporative loss component. In contrast to the effect of vapor pressure modification upon the amount of material emitted, the alternative front-end composition modification has no ap-

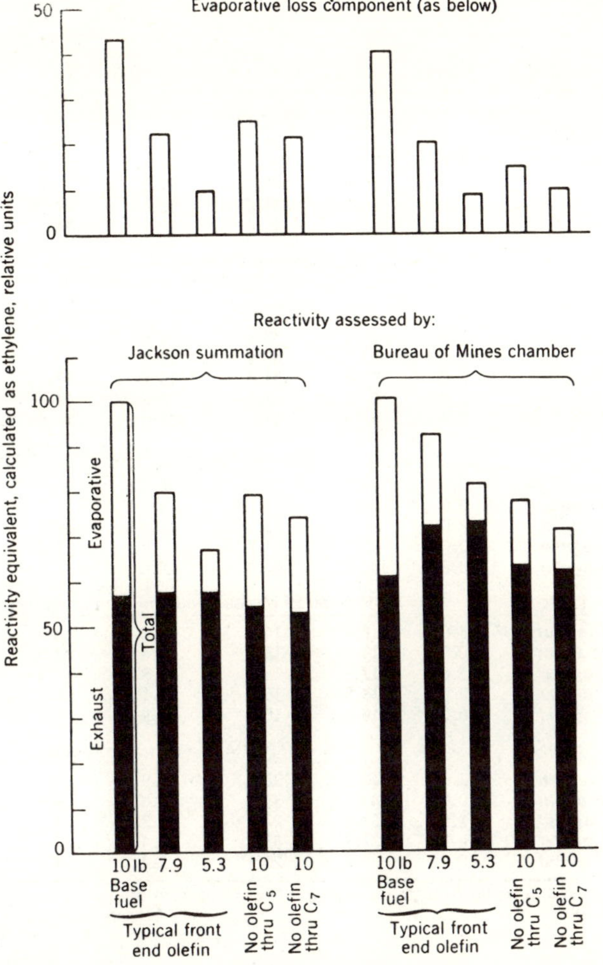

Figure 6. Reactivity measurement by different criteria.

preciable effect upon the amount of emission. Interestingly, it appears that vapor pressure reduction does have an adverse effect upon the amount of hydrocarbon discharged in the exhaust. This effect is seen in the RVP reductions to 7.9 and 5.3 lb, respectively, resulting in emissions that are increased over emissions of the base fuel. These increases amount to about 8%.

Quite a different relationship of fuel composition to emissions is seen when the fuel effect is assessed with reference to photochemical behavior rather than with reference to gross hydrocarbon. As illustrated in Figure 5, volatility reduction within the range of 10 to 5.3 lb would reduce the photochemical effect by about 20% maximum in contrast to the nearly 30% indicated by a gross hydrocarbon measurement.

In comparison with the 20% reduction in reactivity of total vehicular emissions that is available through reduction of RVP from 10 to 5.3 lb, about 27% reduction is available through modification of the front-end fuel composition. This maximum benefit via

Table V. Comparison of emissions from leaded and nonleaded fuels.

Code No.	Grade	Lead ml/gal	Average of 70° and 95°F Tests[a]				
			Total	Exhaust	Carburetor	Tank	Carburetor + Tank
			Gross Hydrocarbon Emissions, g per test				
1	Premium	2.6	87.2	44.6	13.0	29.6	42.6
8	Premium	0.0	87.9	47.9	11.6	28.4	40.0
7	Regular	2.0	78.9	47.3	11.6	20.0	31.6
10	Regular	0.0	84.2	47.8	12.4	24.0	36.4
11	Regular	0.0	81.8	46.6	13.4	21.8	35.2
			Hydrocarbon Emissions — Reactivity Equivalent, Ethylene, g per test				
			Bureau of Mines Chamber Evaluation				
1	Premium	2.6	70.3	38.8	9.4	22.1	31.5
8	Premium	0.0	83.1	49.9	10.0	23.2	33.2
7	Regular	2.0	62.0	34.0	10.7	17.3	28.0
10	Regular	0.0	69.2	47.0	8.0	14.2	22.2
11	Regular	0.0	81.2	41.2	16.0	24.0	40.0
			GM NO_2 Formation Scale by Jackson				
1	Premium	2.6	40.2	21.4	5.8	13.0	18.8
8	Premium	0.0	40.8	22.0	6.0	12.8	18.8
7	Regular	2.0	39.9	23.4	6.4	10.1	16.5
10	Regular	0.0	36.2	22.2	5.0	9.0	14.0
11	Regular	0.0	45.2	24.2	8.6	12.4	21.0
			Bureau of Mines Eye Irritation (Composite)[b] Scale				
1	Premium	2.6	18.3	14.4	1.3	2.6	3.9
8	Premium	0.0	19.6	15.7	1.4	2.5	3.9
7	Regular	2.0	17.9	14.7	1.3	1.9	3.2
10	Regular	0.0	18.5	15.7	1.1	1.7	2.8
11	Regular	0.0	19.5	15.1	1.9	2.5	4.4
			GM Eye Irritation[b] Scale				
1	Premium	2.6	40.3	32.2	2.9	5.2	8.1
8	Premium	0.0	40.7	33.1	2.8	4.8	7.6
7	Regular	2.0	32.6	27.2	2.3	3.1	5.4
10	Regular	0.0	37.1	31.4	2.4	3.3	5.7
11	Regular	0.0	36.8	29.1	3.4	4.3	7.7

[a] Each fuel was used in replicate tests on eight vehicles at each of the two temperatures.
[b] Definitions in Appendix.

front-end modification is realized in replacing C_4 through C_7 olefin with saturated material. A less drastic modification to the fuel is shown by the data of Table III to yield about 23% reduction in the photochemical effect.

The implication arising from comparing the two assessment criteria — namely, quantity of emissions and photochemical effect — is that the method of assessing the effect of fuel modification has critical influence upon the conclusion that will be reached. If the photochemical effect of emissions is of primary concern then it would appear that direct experimental measurement of the effect patently is preferable to other means of emission measurement.

The data illustrated in Figure 5 reflect fuel assessments by experimental irradiation chamber measurements. An alternative method of reactivity assessment permits calculation of reactivity without making any experimental measurements. The two methods, direct measurement and application of reactivity scale, may yield significantly different values for emission reactivities. The Jackson scale, or summation data, are given in Table III and illustrated in Figure 6 to permit comparison with the experimental data

Table VI. Change in emission characteristics with change from leaded to nonleaded fuel.

Grade	Fuel Change	Change in Emission Characteristics with Indicated Fuel Change (The leaded fuel is the reference fuel)		
		Specific Reactivity—R_{NO_2}—BuMines Chamber		
		Exhaust, %	Evaporative, %	Total, %
Premium	from Leaded to Clear[a]	+20	+12	+17
Regular	from Leaded to Low-olefin Clear	+36	—32	+4
Regular	from Leaded to High-olefin Clear	+22	+28	+25
		Reactivity Equivalent—R_{NO_2}—BuMines Chamber		
		Exhaust, %	Evaporative, %	Total, %
Premium	from Leaded to Clear	+29	+12	+21
Regular	from Leaded to Low-olefin Clear	+38	—31	+7
Regular	from Leaded to High-olefin Clear	+21	+28	+24
		Reactivity Equivalent—Eye Irritation—BuMines		
		Exhaust, %	Evaporative, %	Total, %
Premium	from Leaded to Clear	+9	+13[b]	+10
Regular	from Leaded to Low-olefin Clear	+7	—25[b]	No change
Regular	from Leaded to High-olefin Clear	+3	+25[b]	+7
		Reactivity Equivalent—Eye Irritation—GM		
		Exhaust, %	Evaporative, %	Total, %
Premium	from Leaded to Clear	+3	—1[b]	+1
Regular	from Leaded to Low-olefin Clear	+15	—7[b]	+11
Regular	from Leaded to High-olefin Clear	+7	+30[b]	+10

[a] Nonleaded.
[b] Absolute values upon which these percentages are based range between about 2 and 10—the associated percentage values are significant only in indicating direction and general order of magnitude.

from chamber reactivity observations. The following differences are noted in an examination of the data and the illustration:

1. The chamber measurement reveals a significantly greater contribution from the exhaust than is indicated by application of the linear summation reactivity scale.
2. In the cases of olefin replacement, chamber measurement indicates greater improvement, i.e., reduction in evaporative emission reactivity than would be estimated by applying the Jackson scale.
3. The net effect considering both exhaust and evaporative losses is that the chamber measurement relative to the reactivity scale calculation appreciates front-end olefin saturation and tends to depreciate the apparent advantage in simply reducing the amount of evaporative loss through volatility reduction.

Effect of fuel modification to maintain octane quality without use of TEL. The experimental data that were obtained in tests with comparative leaded and nonleaded fuels are given in Table V. The gross hydrocarbon measurements reveal small but significant differences in the evaporative losses and it is relevant to question whether these differences are due to chemical differences, per se, or to volatility differences. To answer the question, the evaporative losses were set against volatility parameters and it was found that the weight differences were assignable to the physical rather than to chemical property. Therefore, to analyze the data for the effect of chemical differences, the evaporative loss (i.e., gross hydrocarbon loss) for each nonleaded fuel was normalized to the loss for the appropriate leaded reference fuel. Using such normalized data, the effect of fuel composition on emission characteristics is shown by data of Table VI. Results from the work with these fuels are properly used comparing only the premium with premium, and regular with regular fuels. The data reflect significant differences between the leaded and nonleaded premium fuels and also show differences of consequence between the emissions from the three regular-grade fuels.

In the case of the premium fuels, BuMines reactivity data reveal that emissions produced using the nonleaded premium exhibit photochemical reactivity that is increased over that of the leaded fuel in both the exhaust and evaporative components. However, the larger effect is upon the exhaust. In this case it would appear, clearly, that the increased photochemical reactivity is attributable to the higher aromatic content of the nonleaded fuel. Of the regular nonleaded fuels only the high-olefin fuel produces emissions with significantly higher reactivity equivalent than the base leaded fuel. However, there were differences between the base leaded fuel and the low-olefin nonleaded fuel wherein the exhaust and evaporative emissions were considered separately. The low-olefin nonleaded fuel yielded a markedly lower reactivity equivalent for the evaporative emissions, but the reduced reactivity equivalent for the evaporative component was largely offset by an increase in the exhaust component. The net effect was that the low-olefin nonleaded fuel was associated with about the same overall photochemical reactivity as the base leaded fuel. Clearly, the high-olefin nonleaded regular was responsible for higher reactivity of emissions overall and for markedly higher reactivity of evaporative emissions.

It would appear from close examination of these data that exhaust reactivity is primarily sensitive to the aromatic content of the fuel, whereas the evaporative component reflects much more strongly the olefinic content.

Using calculated, i.e., linear summation, reactivity data, the issue of leaded versus nonleaded fuel is pictured differently. Reactivity data by the Jackson NO_2-formation scale reflect much lesser differences between leaded and nonleaded fuels in both the premium and regular categories. By this criterion, exhaust reactivity is indicated to be insensitive to either the aromatic or the olefin content of the fuel, whereas evaporative emission reactivity reflects — though to a small degree — the olefinic content.

Reactivity data by the Bureau of Mines eye-irritation (composite) scale reveal a fuel-to-emission correspondence similar to the one based on Jackson data. Interestingly, use of the GM eye-irritation reactivity data did not yield a markedly different picture insofar as effect of fuel aromatics on emission reactivity is concerned. This lack of a stronger effect of fuel aromatics is surprising considering that, by the GM eye-irritation scale, aromatics have very high ratings, and that the aromatic content of fuel is mirrored in the exhaust emissions.

While the finding was surprising it is explained in terms of the composition products that were found. Exhaust from the two fuels contained approximately equal concentrations of the highly irritant toluene that accounts for a large percentage of the aromatic contribution to eye irritation. Thus the difference in aromatic between the two exhausts is to be found in the heavy aromatic that, as eye irritant, is comparable to olefin. Differences in the nonaromatic can then offset differences in aromatic, and the less aromatic fuel did produce emissions richer in olefin by about 3%. Other comparable compositional differences serve partially to offset the influence of the higher aromatic content of the one exhaust to result in roughly comparable eye irritation for the two.

Summary

Volatility and composition characteristics of fuels were evaluated as factors affecting the photochemical smog potential of automotive emissions; the study was designed to provide information of direct utility in predicting or assessing the comparative benefits to be derived from different fuel modifications. For use in assessing reactivity of emissions, a correlation method was developed to provide estimates of reactivity as manifested in large irradiation chambers under simulated atmospheric conditions. Results from these experimental reactivity measurements differed significantly from results obtained by computing emission reactivity solely on the basis of individual hydrocarbon reactivity assignments (i.e., using reactivity scales). The differences are attributed to inadequacy of the computational (reactivity scale) technique to recognize component interaction or to provide any input for reactive components actually in the exhaust but not recognized in the hydrocarbon analysis.

Relative benefits of different fuel modifications were studied utilizing experimental measurement of photochemical behavior. The photochemical smog potential of vehicle emissions considering both evaporative and exhaust emissions was found to be significantly reduced by either reducing fuel volatility or by replacing light olefinic with saturated material. Changes in fuel formula to provide adequate octane quality without use of lead additive were found to affect the pollution potential of the emissions in significant degree and the effects were largely unfavorable.

Acknowledgment

In the execution of this experimental program technical liaison to API and assistance both in program design and in data interpretation were provided by an API Task Force — Chairman, Mr. R. K. Stone, Chevron Research Company. Other members of the committee were Dr. R. S. Spindt, Gulf Research and Development Co.; Mr. H. E. Alquist, Phillips Petroleum Co.; Dr. J. E. Gerrard, Esso Research and Engineering Co.; and Mr. J. B. Duckworth, American Oil Company.

Fuels were selected in collaboration with the API Fuels Technology Task Force: S. S. Sorem, Shell Oil Company, Chairman.

The program was conducted as a part of a larger study organized and given oversight by the Engine Fuels Subcommittee (J. B. Rather, Jr., Mobil Oil Corporation, Chairman) of the API Engineering and Technical Research Committee (L. A. McReynolds, Phillips Petroleum Company, Chairman).

References

1. Eccleston, B. H., Noble, B. F., and Hurn, R. W., "Influence of volatile fuel components on vehicle emissions," U. S. Dept. of the Interior, Bureau of Mines, Report of Investigations No. 7291 (1969) (in press).
2. Spindt, R. S., and Sorem, S. S., "Effect on vehicle emissions of API prototype lead-free fuels," API Preprint No. 42-69, presented at the 34th Midyear Meeting of API, Palmer House, Chicago, Ill., May 13, 1969.
3. Stone, R. K., and Eccleston, B. H., "Vehicle emissions vs. fuel composition," API Preprint No. 41-69, presented at the 34th Midyear Meeting of API, Palmer House, Chicago, Ill., May 13, 1969.
4. Fleming, R. D., Dimitriades, B., and Hurn, R. W., "Procedure in sampling and handling auto exhaust," *J. APCA* **15** (8): 371–374 (August 1965).
5. Seizinger, Donald E., "High resolution gas chromatographic analysis of auto exhaust gas," Perkin-Elmer *Instrument News*, **18** (1): 11–12 (October 1967).
6. Dimitriades, Basil, "Methodology in air pollution studies using irradiation chambers," *J. APCA*, **17** (7): 460–466.
7. Dimitriades, Basil, "Recent findings concerning effects of extraneous factors on hydrocarbon reactivity measurements using irradiation chamber," presented at the 8th Conference on Methods in Air Pollution & Industrial Hygiene Studies, Oakland, Calif., February 6–8, 1967.
8. Jackson, Marvin W., "Effects of some engine variables and control systems on composition and reactivity of exhaust hydrocarbons," *SAE Vehicle*

Table VII. Eye irritation reactivity scales.

Component(s)	Altshuller Scale	GM Scale	Composite scale[a]	
			Absolute rating	Ethylene equivalents
C_1-C_5 Paraffins	0	0	0	0
C_6+ Paraffins	0	0.5	0.5[b]	0.17
Acetylene	0	0	0	0
Ethylene	5	1	3	1
Propylene	6	3.9	5	1.7
1-Alkenes	6	2.5	4.3	1.4
Internal alkenes	6	1.8	3.9	1.3
Diolefins	10	6.9	8.5	2.8
Benzene	0	1	1	0.3
Monoalkylbenzenes	4	(3.3)[c]	(3.6)[c]	(1.2)[c]
Benzyls	4	5.3	4.7	1.6
non-benzyls	4	1.3	2.7	0.9
Polyalkylbenzenes	6	2.7	4.4	1.5
Aromatic olefins	(10)[c]	8.3	(9.1)[c]	(3.0)[c]
Aliphatic aldehydes	(5)[c]	(2.0)[c]	(3.5)[c]	(1.2)[c]
Aromatic aldehydes	(8)[c]	(8)[c]	(8)[c]	(2.7)[c]

[a] Composite scale values were obtained by averaging the Altshuller and GM scale values except when footnoted otherwise.
[b] GM reactivities are believed to be more accurate because they were determined from 6-hr irradiation tests rather than from 2-hr tests or from dynamic tests; therefore, GM values were used in the composite scale.
[c] Values are suggested by authors.

Emissions, Part II, V. **12**: 241–267 (covers 1963–1966).

9. Glasson, William A., and Tuesday, Charles S., "Inhibition of the atmospheric photooxidation of hydrocarbons by nitric oxide," *GMR-475*, General Motors Corporation (March 31, 1965).

Appendix

Eye-Irritation Reactivity Scales for Hydrocarbons

Two reactivity scales based on eye irritation can be found in the literature. One has been proposed by Altshuller* and is based on data generated by a number of studies; the other is based on a recent study at GM laboratories.† Data of Table VII show the correspondence of hydrocarbon ratings by the two scales.

These two scales are different to an extent that conclusions from studies may be significantly different depending on which scale was used. Therefore, the two scales cannot be used indistinguishably and adoption of one for use must be justified. Deliberations regarding appropriateness of the two scales were based on consideration of the following relative merits and shortcomings:

Altshuller Scale

1. Data were obtained from a number of studies where experimental conditions such as reactant concentrations, irradiation times, and apparatus varied widely. Therefore, validity of resultant rating is wider in principle. However, some uncertainty is likely to characterize relative ratings, because of conjecture that unavoidably enters in reviewing and correlating results from widely different studies.

2. The zero ratings for the unreactive paraffins and benzene are probably unrealistic. This is because these ratings were based on dynamic irradiation tests or tests with short irradiation times. Such tests are terminated long before completion of NO photooxidation; therefore, under these conditions, the tested hydro-

---

* Altshuller, A. P., "An evaluation of techniques for the determination of the photochemical reactivity of organic emissions," *Jour. of APCA,* **16**: 5, 257–260 (May 1966).

† Heus, Jon M. and Glasson, William A., "Hydrocarbon reactivity and eye irritation," *Environmental Science & Tech.,* **2**: 12, 1109–1116 (December 1968).

carbon is not given an opportunity to develop its eye-irritation potential and resultant reactivity is erroneously taken to be zero.
3. Ratings are based on data from old studies in which chamber methodology was undeveloped yet; therefore, some inaccuracies in relative ratings should be expected.

GM Scale

1. Internal consistency of relative ratings should be high because data are based on a single study with one set of experimental conditions. However, for this same reason the validity of the scale should be limited.
2. Relative ratings for highly reactive hydrocarbons are probably erroneoulsy low because in the respective tests the hydrocarbon reactant may have been consumed long before the eye-irritation measurement was made; under such conditions the measurement result does not reflect the hydrocarbon's maximum yield in eye irritants.

In conclusion, it does not seem possible to declare one or the other of the two scales as the "correct" scale. Therefore, it was resolved to combine the two scales into a composite one in which the ratings were equal to either the averages of those by the two scales or to those by the GM scale. Resultant ratings for composite scale and explanations regarding their derivation are given in Table VII. Because the two scales have approximately the same rating span, namely $0 - 10$, averaging was applied on rating values as reported, rather than after conversion into ethylene-equivalent values. The composite scale ratings were then divided by that of ethylene to express each component's eye-irritation reactivity in terms of ethylene equivalents.

The Oxides of Nitrogen and their Detection in Automotive Exhaust

John H. McFarland
and C. S. Benton

The oxides of nitrogen are receiving more and more attention as air pollutants. They are formed from a variety of sources, almost all involving combustion of fossil fuels (see Table 1). Nitrogen oxides are important because they participate in many photochemical reactions and are largely responsible for the formation of the "brown haze" of smog and the accompanying eye irritation and other physiological effects.

Some physical data on the important oxides of nitrogen are listed in Table 2. Of these oxides, only nitrogen oxide (NO) and nitrogen dioxide (NO_2), together referred to as NO_x, are of concern in air pollution studies. Nitrous oxide (N_2O) has no effect at low concentrations and emission sources are rare. Dinitrogen tetraoxide $(NO_2)_2$ or N_2O_4 is a polymer of NO_2 which forms when NO_2 is compressed and liquefied.

$$2NO_2 \rightleftarrows N_2O_4$$

At the concentrations usually encountered in the atmosphere, equilibrium is very quickly attained and the quantity of N_2O_4 is so small as to be of minimal importance in air pollution studies.

NO is present in automobile exhaust in concentrations ranging from a few parts per million to several thousand ppm. It is a colorless, odorless gas. Little is known about the toxicological effects of NO since at high concentrations NO converts into NO_2. During the combustion of fossil fuels, the nitrogen and oxygen in the air used combine at high temperatures to form NO.

$$N_2 + O_2 \rightleftarrows 2NO$$

The equilibrium concentration of the reaction depends on the flame temperature at which combustion occurs, the length of time the combustion gases are maintained at that temperature, and the amount of excess air present in the flame. Once NO is formed the decomposition rate is too slow under the ordinary conditions of rapid cooling to allow the NO to dissociate into oxygen and nitrogen. Thus the NO is "frozen" in the combustion products on leaving the high-temperature zone. In automotive exhaust gases at complete combustion and top speed the concentration of NO may be as high as 4000 ppm (see Table 3).

While most of the NO_x that is emitted in automobile exhaust is in the form of NO, nitrogen dioxide begins to form by the reaction of the NO with excess oxygen

$$2NO + O_2 \rightleftarrows 2NO_2$$

The reaction proceeds more slowly as the concentration of NO decreases so that at low concentrations, NO levels may be stable for a long time. It has been shown that ultraviolet photolysis affects the conversion of NO to NO_2 (4). Also a variety of hydrocarbons are known to catalyze the reaction (5). Therefore, most investigations lump NO and NO_2 together as NO_x for determination purposes since at most concentrations and conditions they are easily interconverted in the atmosphere. The ratio of NO to NO_2 varies with the time of day (decreasing during the day and increasing at night), depending on the action of sunlight, oxygen, and other oxidizing or reducing agents present. NO_2 is a brown pungent gas. Concentrations of 20–50 ppm are irritable to the eyes. At only 150 ppm there is danger of strong local irritations especially to the respiratory organs (3).

Table 1. Total Estimated Oxides of Nitrogen Emissions from Mobile and Stationary Sources in United States, 1968 (1)

Fuel type	NO_x estimated tons, 1968	% total
Gas and LPG	4,640,000	22.4
Petroleum	9,350,000	45.2
Auto/gasoline	6,600,000	31.9
Coal	4,000,000	19.3
Coal utilities	2,970,000	14.3
Waste burning	746,000	3.6
Coal wastes	190,000	0.9
Industrial non-combustion	200,000	1.0
Agricultural, wood, and forest fires	1,733,000	8.4
Total	20,699,000	

Table 2. Properties of the Oxides of Nitrogen (2)

	Specific gravity	Solubility in 100 pts H_2O	ROH	mp, °C	bp, °C
Nitric oxide NO or $(NO)_2$	1.0367^a	7.34 ml	26.6 ml	−161	−151
Nitrogen Dioxide NO_2 or $(NO_2)_2$	1.448^{20}		decomposes	−9.3	21.3
Nitrous oxide N_2O	1.530^a	130.52 ml	soluble	−102.3	−90.7
NO and NO_2 are paramagnetic					

[a] With reference to air = 1.

Table 3. Composition of Automobile Exhaust Gases (3)

	Idling	Full load Low Speed	High Speed
Nitrogen oxides	0–50 ppm	1000 ppm	4000 ppm
CO_2	6.5–8 vol%	7–11%	12–13%
H_2O	7–10 vol%	9–11%	10–11%
O_2	1–1.5 vol%	0.5–2%	0.1–0.4%
CO	3–10 vol%	3–8%	1–5%
H_2	0.5–4 vol%	0.2–1%	0.1–0.2%
Hydrocarbons	300–8000 ppm	200–500 ppm	100–300 ppm

Lead compounds (Pb) 60 mg/m³, 3,4-Benzopyrene 0.001–0.01 mg/m³

It is assumed that nitrogen and small amounts of inert gases make up the remaining volume

Numerous wet chemical and instrumental methods have been reported in the literature which measure the oxides of nitrogen. In the remainder of this paper I would like to briefly describe some of those methods.

Acidimetric Determination (3)

Acidimetric determination of NO_x is common in industrial determination. The gas sample is introduced into a large evacuated flask. Water or dilute hydrogen peroxide is added and the mixture is allowed to stand for several hours before being titrated to a methyl red end point with NaOH. Reactions are as follows

$$3H_2O_2 + 2NO \rightarrow 2HNO_3 + 2H_2O$$

$$NaOH + HNO_3 \rightarrow H_2O + NaNO_3$$

The waiting time can be cut down by shaking the sample and hydrogen peroxide vigorously with a neutral foaming agent for 5 min.

Phenoldisulfonic Acid Method (3, 6, 7)

A sample is collected in an evacuated flask containing a dilute solution of sulfuric acid and hydrogen peroxide. All nitrogen oxides in the sample are oxidized to nitric acid. The resultant solution is made just basic with 1 N sodium hydroxide and evaporated to dryness. Phenoldisulfonic acid reagent and ammonium hydroxide are added to form the yellow trialkali salt of 6-nitro-1-phenol-1,4-disulfonic acid. A colorimetric determination is then made at 410 mμ and compared to a calibration curve plotted for known amounts of $NaNO_3$ treated the same way. The main drawback to this method is the fact that it is very tedious and time consuming. It is reproducible and relatively accurate, but it is subject to interference from sulfur dioxide, halogens, inorganic nitrates, and organic nitrogen compounds.

Dry Absorption of NO_x

According to Peters and Straschil (8), NO_x can be determined by absorption on sodium chlorite ($NaClO_2$). The sample is drawn through a glass tube containing a dried paste of alumina and $NaClO_2$. The contents of the tube are then heated with NaOH until the particles decompose. They are then dissolved in water and reduced with Devarda alloy to ammonia. The ammonia formed is photometrically determined. The reactions are presumed to be

$$NO + NaClO_2 \rightleftarrows NaNO_3 + NaCl$$

$$5OH^- + 3NO_3^- + 8Al^0 + 2H_2O \rightleftarrows 8AlO_2^- + 3NH_3$$

p-Anisidine Test Paper

A test paper for NO_2 has been described by Gelman, *et al.* (*9*). A rubber bulb hand aspirator is used to draw a known volume of air through a test paper treated with *p*-anisidine. Any red-brown color produced is compared with a series of standard colors. Although the method is easy to operate and gives good color differentiation in the range of 2–10 ppm, it has two disadvantages. The test paper is not stable for more than two weeks, and the response is dependent on atmospheric humidity. These problems were partially alleviated by Christie, *et al.* (10), who added glycerol as a humectant during preparation of the test papers. Storage life was increased to 3 mon and the effect of humidity at normal levels was minimized.

Determination by Photometry of Azo Dyes

This method has been used and modified by a great number of investigators. NO_2 is first absorbed as nitrite. Then an aromatic amine is used to form a diazonium compound. A second aromatic amine couples with the diazonium compound formed to give an azo dye which is then colorimetrically determined. The classic Griess reagent is a mixture of sulfanilic acid and α-naphthylamine in acetic acid. The reactions, as believed to occur in the Griess determination, are as follows

$$2NO_2 + H_2O \longrightarrow HNO_3 + HNO_2$$

A known volume of sample gas is bubbled into the reagent through a fritted scrubber. After a waiting period the absorbance is measured at 530 mμ and compared to a standard curve plotted from solutions of $NaNO_2$. The NO_2 to NO_2^- ratio is 2 to 1.

There are many modifications of the Griess determination. Total oxides of nitrogen can be measured by first oxidizing NO to NO_2. NO could then be determined by subtracting NO_2 from NO_x after parallel determinations of NO_2 and NO_x. Several oxidants have been used (*3*) including $KMnO_4$ in sulfuric acid, CrO_3 on dry surfaces, uv irradiation in a long quartz spiral (which can be catalytically accelerated by butadiene) and ozone.

A modification by Jacobs and Hochheiser (*11*) uses an alkaline solution to absorb NO_2 according to the equation

$$2NO_2 + 2NaOH \rightleftharpoons NaNO_2 + NaNO_3 + H_2O$$

Only the nitrite formed participates in the diazotization. Interfering SO_2 can be eliminated by adding a drop of H_2O_2. However, it is difficult where sufficient SO_2 has accumulated for the determination.

The most popular modification used is the Saltzman method (*12*). Saltzman's reagent uses N-(1-naphthyl)-ethylenediamine hydrochloride instead of α-naphthylamine. The absorption is measured at 550 mμ. The reactions are the same as those given except for the α-naphthylamine.

From the equations it is expected that 0.5 moles of nitrate is equivalent to 1.0 moles of NO_2 in color intensity. However, Saltzman reported a factor of 0.72. There has been some controversy over this factor. Stratman and Buck report that the factor is unity (*13*). Others report factors similar to Saltzman's factor (*14, 15*).

There are a number of modifications on the Saltzman method. One reported by Lyshkow (*16*) improves the rate of color development, sensitivity and longer shelf-life of the reagent by adding a promoter (R-salt), and optimizing the concentrations of diazotizing and coupling reagents. The R-salt used was 2-napthol-3,6-disulfonate. Another promising modification is the syringe method of Newmark (*14*). This method uses a 50-ml hypodermic syringe to collect the sample and greatly simplifies the procedure.

Spectrophotometric Determination

Direct determination of NO_x is possible at high concentrations or over long distances. One such determination was made with a path length of 3 km using a search light as energy source to measure atmosphere concentrations (*3*). More recently several systems (example, Fig. 1) have been described which measure NO_x concentration in the 350–450 mμ region where there is little interference with other exhaust gases (*17–20*). Since most of the NO_x coming out of the exhaust pipe is in the form of

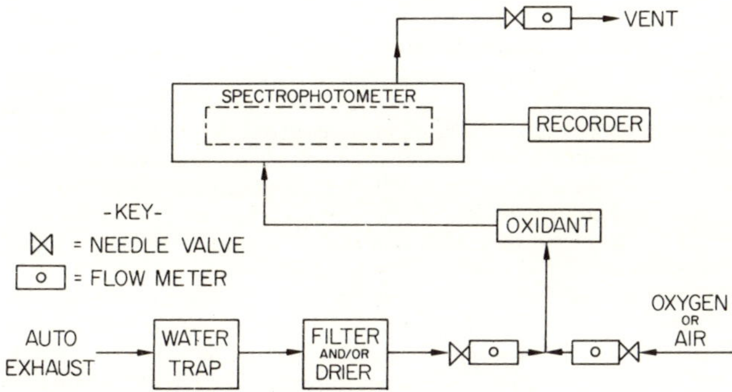

Figure 1. Exhaust gas sampling and dilution system.

NO, which does not absorb, an oxidation step is required. Various oxidants have been used, including ozone, chlorine dioxide, and permanganate. In most cases the exhaust gas was passed through a water trap and a filter, mixed with oxygen, compressed and oxidized and then measured. Ripley, *et al.* (*19*), report excellent conversion using glass fiber paper saturated with a solution of sodium dichromate and sulfuric acid. This removes a relatively long residence time needed to achieve complete oxidation. Also NO_2 may be lost during that time due to reactions with other exhaust gases.

Nitrate-Specific Ion Electrode (*21*)

By the following series of reactions NO and NO_2 can be oxidized and hydrolyzed to get nitrate ion

$$NO + O_3 \rightleftarrows NO_2 + O_2$$

$$2NO_2 + O_3 \rightleftarrows N_2O_5 + O_2$$

$$N_2O_5 + H_2O \rightleftarrows 2HNO_3$$

$$3N_2O_4 + 2H_2O \rightleftarrows 4NO_3^- + 4H^+ + 2NO$$

The nitrate ion is then quantitatively determined using a nitrate selective ion electrode. Electrode response is relatively stable. There are few interferences since most trace gases in the atmosphere do not hydrolyze to form ions which cause specific ion electrode interference. Those that do are in negligible concentrations.

Homogeneous Chemiluminescence (22, 23)

The homogeneous gas phase reaction of NO and NO_2 with ozone or atomic oxygen results in light emission. NO may be determined by its reaction with ozone

$$NO + O_3 \rightleftharpoons NO_2^- + O_2$$

$$NO_2^- \rightleftharpoons NO_2 + h\nu$$

NO_x may be determined by reaction with atomic oxygen according to the equation

$$NO_2 + (O) \rightleftharpoons NO + O_2$$

$$NO + (O) \rightleftharpoons NO_2^-$$

$$NO_2^- \rightarrow NO_2 + h\nu$$

Figure 2 diagrammatically shows the instrument used. The air being monitored for the first reactant gas (NO or NO_x) and a large excess of the second reactant gas (O_3 or O) enter the spherical reaction vessel through separate controlled inlets. A vacuum pump maintains a low pressure which allows the reactant gases to mix rapidly and continuously, emitting light in direct proportion to the concentration of the second reactant gas. The light intensity is sensed by a phototube after a light filter has screened out emissions from unwanted side reactions.

There are advantages to this method. Emissions are specific for the pollutant being monitored. The suitable choice of a light filter and the second reactant should allow interference-free measurements. The chemiluminescent light intensities from homogeneous gas-phase reactions in continuous flow systems are rather insensitive to changes in surface properties.

Gas Chromatography (24, 25)

NO_2 has a radical character and thus exhibits a high affinity for free electrons which allows it to be detected by the electron-

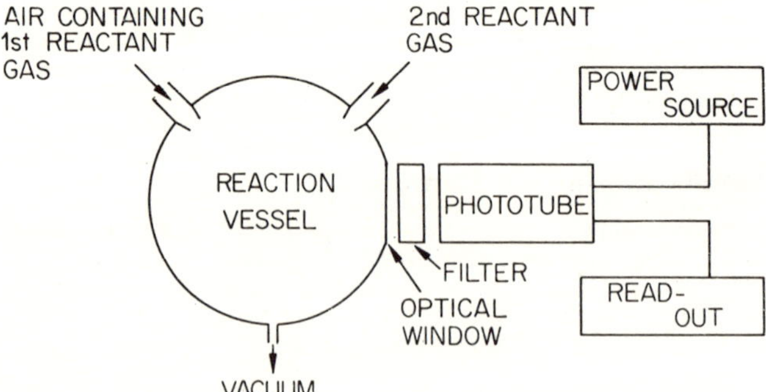

Figure 2. System for analyzing NO by homogeneous chemiluminescence.

capture detector. Morrison and Corcoran (*24*) have described such a detection system that is good in the low ppm range. Other investigators have used thermo conductor and ionization detectors and a great variety of column packings. Problems encountered have included the need for more than one column to separate the various pollutants in air and irreversible absorption on the column packings. Bethea and Meador (*25*) present an extensive review of the literature for columns used for the determination of a number of air pollutants. Their own experimental work indicates that a combination of columns is needed to separate and detect the oxides of nitrogen from other gases.

Sealed Electrochemical Sensors (26)

Quite recently sealed electrochemical sensors have been developed. These sensors put together two electrodes, the sensing electrode and the counter electrode, along with an appropriate electrolyte and a protective semipermeable membrane covering the electrodes and electrolyte. The molecules of the pollutant gas diffuse through the protective membrane and are electro-oxidized or electroreduced (as in the case of NO_2) generating a current directly proportional to the partial pressure of the pollutant gas. Of the major pollutants only sulfur dioxide and the oxides of nitrogen are sufficiently electroactive. Selectivity of NO_2 is obtained by electroreduction in aqueous acid electrolyte because the required potential is higher than that of SO_2.

Chard and Cunningham (*26*) have used sweep voltametric characterization to evaluate high boiling organic electrolyte solutions for the detection of NO_x. Such electrolytes would solve the problems of the inertness of temperatures close to room temperature. The electrolyte, 1,2-propanediol cyclic carbonate-KPF_6 was found to be suitable for the sensing of NO_x at a potential of about 0.70 V on the sensing electrode.

Other

Much attention is being paid to the problem of instrumentation in NO_x detection. Most instrument companies have developed or are in the process of developing quick, reliable, portable NO_x detectors that can be operated at test sites by nontechnical personnel. Such an instrument has been developed by Mast Development Company (*27*). This instrument is based upon a coulometric system with a bromide sensing solution. The reaction is

$$NO_2 + 2H^+ + 2Br^- \rightleftharpoons NO + H_2O + Br_2$$

The instrument is sensitive in the range 0–5000 ppm NO_2.

An instrument developed by Purad, Inc. (*28*) involves the selective oxidation of reactive components over a selective catalyst. Their instrument measures hydrocarbon and carbon monoxide by measuring the temperature rise of a catalyst. This approach might also be adapted for NO_x detection, using any of the selective catalysts developed for NO_x removal.

Literature Cited

(1) HARRIS, R. L., "Control Techniques for Nitrogen Oxides from Stationary Sources," National Air Pollution Control Administration, **1970**.

(2) SHEEHY, J. P., ACHINGER, W. C., AND SIMON, R. A., "Handbook of Air Pollution," Government Publications Office, **1971**, p. 14-12 and 14-13.

(3) LEITHE, W., "The Analysis of Air Pollutants," Ann Arbor-Humphrey Science Publishers, 1970, p. 11 and pp. 177–188.

(4) HALL, H. J., AND BARTOK, W., *Envir. Sci. and Tech.*, **5**, 320 (1971).

(5) STEPHENS, E. R., AND PRICE, M. A., *Atmospheric Envir.*, **3**, 573 (1969).

(6) FISHER, G. E., and CALABRO, D. S., *J. Air Poll. Control Assoc.*, **20**, 666 (1970).

(7) GROTH, R. H., AND CALABRO, D. S., *Journal of the Air Pollution Control Association*, **19**, 884 (1969).

(8) PETERS, K., AND STRASCHIL, H., *Angew. Chem.*, **68**, 291 (1956).

(9) GELMAN, C., GAMSON, R., AND KLAPPER, H., *Proc. 51st Ann. Meeting Air Pollution Control Assoc.*, **19**, (1958).

(10) CHRISTIE, A. A., LINDZEY, R. G., AND RADFORD, D. W. F., *Analyst*, **95**, 519 (1970).

(11) JACOBS, M. B., AND HOCHHEISER, S., *Anal. Chem.*, **30**, 426 (1958).

(12) SALTZMAN, B. E., *Anal. Chem.*, **26**, 1949 (1954).

(13) STRATMANN, J., AND BUCK, M., *Int. J. Air Water Poll.*, **10**, 313 (1966).

(14) NEWMARK, P., "Syringe method for Measuring NO and NO_2 in 200–2000 ppm Range," Paper presented at Pacific Conference on Chemistry and Spectroscopy on November 1, 1967 (available L.A. pollution control lab).

(15) SCARINGELI, F. P., ROSENBERG, E., AND REHME, K. A., *Envir. Sci. and Tech.*, **4**, 924 (1970).

(16) LYSHKOW, N. A., *J. Air Poll. Control Assoc.*, **15**, 481 (1965).

(17) NICKSIC, S. W., AND HARKINS, J., *Anal. Chem.*, **34**, 985 (1962).

(18) SWEENEY, M. P., SWARTZ, D. J. ROST, G. A., AND CHAO, J., *J. Air Poll. Control Assoc.*, **14**, 249 (1964).

(19) RIPLEY, D. C., CLINGENPEEL, J. M., AND HURN, R. W., *Int. J. Air Water Poll.*, **8**, 455 (1964).

(20) KENNAN, C. A., *Anal. Instr.*, **4**, 57 (1966).

(21) DI MARTINI, R., *Anal. Chem.*, **42**, 1102 (1970).

(22) FONTIJIN, A., SABADELL, A. J., AND RONCO, R. J., *Anal. Chem.*, **42**, 575 (1970).

(23) O'KEEFE, A. E., *IEEE Trans. Geoscience Electronics*, **8**, 145 (1970).

(24) MORRISON, M. E., AND CORCORAN, W. H., *Anal. Chem.*, **3a**, 255 (1967).

(25) BETHEA, R. M., AND MEADOR, M. C., *J. Chromatogr. Sci.*, **7**, 655 (1969).

(26) CHAND, R., AND CUNNINGHAM, P. R., *IEEE Trans. Geoscience Electronics*, **8**, 158 (1970).

(27) ROSTENBACH, R. E., AND KLING, R. G., "Nitrogen Dioxide Detection Using a Coulometric Method," Paper presented at the 55th Annual Meeting of APCA, Sheraton-Chicago Hotel, May 20–24, 1962, Chicago, Ill.

(28) HEYLIN, M., *Chem. Eng. News*, **49**, 78 (February 15, 1971).